THE DISTRIBUTION OF PRIME NUMBERS

Large sieves and zero-density theorems

BY

M. N. HUXLEY

OXFORD
AT THE CLARENDON PRESS
1972

Oxford University Press, Ely House, London W. 1

GLASGOW NEW YORK TORONTO MELBOURNE WELLINGTON
CAPE TOWN IBADAN NAIROBI DAR ES SALAAM LUSAKA ADDIS ABABA
DELHI BOMBAY CALCUTTA MADRAS KARACHI LAHORE DACCA
KUALA LUMPUR SINGAPORE HONG KONG TOKYO

Printed in Great Britain
at the University Press, Oxford
by Vivian Ridler
Printer to the University

TO

THE MEMORY OF

PROFESSOR H. DAVENPORT

PREFACE

THIS book has grown out of lectures given at Oxford in 1970 and at University College, Cardiff, intended in each case for graduate students as an introduction to analytic number theory. The lectures were based on Davenport's *Multiplicative Number Theory*, but incorporated simplifications in several proofs, recent work, and other extra material.

Analytic number theory, whilst containing a diversity of results, has one unifying method: that of uniform distribution, mediated by certain sums, which may be exponential sums, character sums, or Dirichlet polynomials, according to the type of uniform distribution required. The study of prime numbers leads to all three. Hopes of elegant asymptotic formulae are dashed by the existence of complex zeros of the Riemann zeta function and of the Dirichlet L-functions. The prime-number theorem depends on the qualitative result that all zeros have real parts less than one. A zero-density theorem is a quantitative result asserting that not many zeros have real parts close to one. In recent years many problems concerning prime numbers have been reduced to that of obtaining a sufficiently strong zero-density theorem.

The first part of this book is introductory in nature; it presents the notions of uniform distribution and of large sieve inequalities. In the second part the theory of the zeta function and L-functions is developed and the prime-number theorem proved. The third part deals with large sieve results and mean-value theorems for L-functions, and these are used in the fourth part to prove the main results. These are the theorem of Bombieri and A. I. Vinogradov on primes in arithmetic progressions, a result on gaps between prime numbers, and I. M. Vinogradov's theorem that every large odd number is a sum of three primes. The treatment is self-contained as far as possible; a few results are quoted from Hardy and Wright (1960) and from Titchmarsh (1951).

Parts of prime-number theory not touched here, such as the problem of the least prime in an arithmetical progression, are treated in Prachar's *Primzahlverteilung* (Springer 1957). Further work on zero-density theorems is to be found in Montgomery (1971), who also gives a wide list of references covering the field.

M. N. H.

Cardiff
1971

CONTENTS

PART I. INTRODUCTORY RESULTS

PART II. THE PRIME-NUMBER THEOREM

PART III. THE NECESSARY TOOLS

PART IV. ZEROS AND PRIME NUMBERS

PART I

Introductory Results

1

ARITHMETICAL FUNCTIONS

An Expotition . . . means a long line of everybody

I. 110

THIS chapter serves as a brief résumé of the elementary theory of prime numbers. A positive integer m can be written uniquely as a product of primes

$$m = p_1^{a_1} p_2^{a_2} \dots p_r^{a_r}, \tag{1.1}$$

where the p_i are primes in increasing order of size, and the a_i are positive integers. We shall reserve the letter p for prime numbers, and write a sum over prime numbers as $\sum_p$ and a product as $\prod_p$. The proof of unique factorization rests on Euclid's algorithm that the highest common factor (m,n) of two integers (not both zero) can be written as

$$(m,n) = mu+nv, \tag{1.2}$$

where u, v are integers. We use (m,n) for the highest common factor and $[m,n]$ for the lowest common multiple of two integers where these are defined.

Let q be a positive integer. Then the statement that *m is congruent to n* (mod q), written $m \equiv n \pmod{q}$, means that $m-n$ is a multiple of q. Congruence mod q is an equivalence relation, dividing the integers into q classes, called *residue classes* mod q. A convenient set of representatives of the residue classes mod q is 0, 1, 2,..., $q-1$. The residue classes mod q form a cyclic group under addition, and the *exponential maps*

$$m \to e_q(am), \tag{1.3}$$

where a is a fixed integer, and

$$e(\alpha) = \exp(2\pi i\alpha), \qquad e_q(\alpha) = \exp(2\pi i\alpha/q), \tag{1.4}$$

are homomorphisms from this group to the group of complex numbers of unit modulus under multiplication. There are q distinct maps, corresponding to $a = 0, 1, 2,\dots, q-1$. They too can be given a group structure, forming a cyclic group of order q. They have the important property

$$\sum_{m \bmod q} e_q(am) = \begin{cases} q & \text{if } a \equiv 0 \pmod{q}, \\ 0 & \text{if not,} \end{cases} \tag{1.5}$$

where the summation is over a complete set of representatives of the residue classes mod q (referred to briefly as a *complete set of residues* mod q). If on the left-hand side of eqn (1.5) we replace m by $m+1$, the sum is still over a complete set of residues, but it has been multiplied by $e_q(a)$, which is not unity unless $a \equiv 0 \pmod{q}$. The sum is therefore zero unless $a \equiv 0 \pmod{q}$, when every term is unity. Interchange of a and m leads to a corresponding identity for the sum of the images of m under a complete set of maps ($a = 0, 1,\dots, q-1$). These identities arise because the images lie in a multiplicative not an additive group.

From Euclid's algorithm comes the Chinese remainder theorem: if m, n are positive integers and $(m, n) = 1$, then any pair of residue classes $a \pmod{m}$ and $b \pmod{n}$ (which are themselves unions of residue classes mod mn) intersect in exactly one class $c \pmod{mn}$, given by

$$c \equiv bmu + anv \pmod{mn} \tag{1.6}$$

in the notation of eqn (1.2). Now let $f(m)$ be the number of solutions (ordered sets $(x_1,\dots, x_r)$ of residue classes) of a set of congruences

$$g_i(x_1,\dots, x_r) \equiv 0 \pmod{m}, \tag{1.7}$$

where the g_i are polynomials in $x_1,\dots, x_r$ with integer coefficients. When $(m, n) = 1$, $g_i(x_1,\dots, x_r)$ is a multiple of mn if and only if it is a multiple both of m and of n. Hence

$$f(mn) = f(m)f(n) \quad \text{whenever } (m, n) = 1. \tag{1.8}$$

Equation (1.8) is the defining property of a *multiplicative arithmetical function.* An *arithmetical function* is an enumerated subset of the complex numbers, that is, a sequence $f(1), f(2),\dots$ of complex numbers. The property

$$f(mn) = f(m)f(n) \tag{1.9}$$

for all positive integers m and n seems more natural; if eqn (1.9) holds as well as (1.8) then $f(m)$ is said to be *totally multiplicative,* but (1.8) is the property fundamental in the theory.

The Chinese remainder theorem enables us to construct more complicated multiplicative functions. We call a residue class $a \pmod q$ *reduced* if the highest common factor (a, q) is unity. A sum over reduced residue classes is distinguished by an asterisk.

With this notation we introduce Euler's function $\varphi(m)$ by

$$\varphi(m) = \sum_{a \bmod m}^{*} 1. \tag{1.10}$$

To show that $\varphi(m)$ is multiplicative, we must verify that in eqn (1.6) $(c, mn) = 1$ if and only if both (a, m) and (b, n) are unity. Equation (1.6) implies also that Ramanujan's sum

$$c_q(m) = \sum_{a \bmod q}^{*} e_q(am) \tag{1.11}$$

is multiplicative in q for each m. We see this if we write

$$a = a_2 q_1 u_2 + a_1 q_2 u_1, \tag{1.12}$$

where

$$q_1 u_2 + q_2 u_1 = 1; \tag{1.13}$$

note that $u_1 \pmod{q_1}$ and $u_2 \pmod{q_2}$ are reduced residue classes, that

$$c_{q_1 q_2}(m) = \sum_{a_1 \bmod q_1}^{*} \sum_{a_2 \bmod q_2}^{*} e_{q_2}(a_2 u_2) e_{q_1}(a_1 u_1), \tag{1.14}$$

and that $a_1 u_1$ runs through a complete set of reduced residues mod q_1 when a_1 does so.

Two examples follow of totally multiplicative arithmetical functions. The first is

$$f(m) = m^s, \tag{1.15}$$

where s is a complex variable

$$s = \sigma + it, \tag{1.16}$$

σ and t being real. This notation is traditional among number theorists.

To introduce our second example we note that the reduced residues mod q (algebraically the invertible elements in the ring of integers mod q) form under multiplication an Abelian group of order $\varphi(q)$. By considering the images of the generators of this group, we can see that from this group to the group of complex numbers of unit modulus under multiplication there are $\varphi(q)$ maps χ with the homomorphism property

$$\chi(mn) = \chi(m)\chi(n). \tag{1.17}$$

These include the *trivial map* for which $\chi(m) = 1$ for each reduced class m. We turn these maps into arithmetical functions by defining

$$\chi(m) = 0 \quad \text{if } (m, q) > 1. \tag{1.18}$$

With this definition, eqn (1.17) still holds. We have now assigned a complex number to each residue class mod q. Hence we have constructed a totally multiplicative periodic function, which is called a *Dirichlet's character* mod q, or more briefly a *character*. Characters can be defined as those totally multiplicative functions that are periodic. Since negative integers also belong to well-defined residue classes mod q, we can speak of $\chi(m)$ when m is a negative integer; in particular, we shall refer to $\chi(-1)$.

It is possible to build new multiplicative functions from old. We say that *d divides m*, written $d \mid m$, when the integer m is a multiple of the positive integer d; another paraphrase is 'd is a *divisor* of m'. (Note that the divisors of -6 are 1, 2, 3, 6.) Now let $f(m)$ and $g(m)$ be multiplicative. Then so are the arithmetical functions

$$h(m) = f(m)g(m), \tag{1.19}$$

$$h(m) = \sum_{d|m} f(d), \tag{1.20}$$

and

$$h(m) = \sum_{d|m} f(d)g(m/d). \tag{1.21}$$

We shall consider eqn (1.21), since (1.20) is a special case, and (1.19) is evident. When $(m, n) = 1$, the divisor d of mn can be written uniquely as $d = ab$, where $a \mid m$ and $b \mid n$, and $(a, b) = 1$. Hence

$$\begin{aligned} h(mn) &= \sum_{d|mn} f(d)g(mn/d) \\ &= \sum_{a|m} \sum_{b|n} f(ab)g(mn/ab) \\ &= \sum_{a|m} \sum_{b|n} f(a)f(b)g(m/a)g(n/b), \end{aligned} \tag{1.22}$$

which is $h(m)h(n)$ as required. Thus

$$d(m) = \sum_{d|n} 1, \tag{1.23}$$

the number of divisors of m, and

$$\sigma(m) = \sum_{d|m} d, \tag{1.24}$$

the sum of the divisors of m, are multiplicative functions.

We can invert eqn (1.20) and return from $h(m)$ to $f(d)$ by using *Möbius's multiplicative function* $\mu(m)$, defined by

$$\left.\begin{aligned} \mu(1) &= 1 \\ \mu(p) &= -1 \quad \text{for primes } p \\ \mu(p^a) &= 0 \qquad \text{for prime powers } p^a \text{ with } a > 1 \end{aligned}\right\}. \tag{1.25}$$

If the positive integer m factorizes according to (1.1), then

$$\sum_{d|m} \mu(d) = \prod_i \{1+\mu(p_i)+\mu(p_i^2)+...+\mu(p_i^{a_i})\}$$
$$= \prod_i (1-1) = 0, \qquad (1.26)$$

unless $m = 1$, when the product in eqns (1.1) and (1.26) is empty. We have now proved the following lemma.

Lemma. *If m is a positive integer, then*

$$\sum_{d|m} \mu(d) = \begin{cases} 1 & \text{if } m = 1, \\ 0 & \text{if } m > 1. \end{cases} \qquad (1.27)$$

From the lemma we have the corollary:

Corollary. *If $h(m)$ and $f(m)$ are related by eqn* (1.20), *then*

$$f(n) = \sum_{m|n} \mu(m)h(n/m), \qquad (1.28)$$

and if eqn (1.28) *holds then so does eqn* (1.20).

To prove the corollary we substitute as follows.

$$\sum_{m|n} \mu(m)h(n/m) = \sum_{m|n} \mu(m) \sum_{d|(n/m)} f(d)$$
$$= \sum_{d|n} f(d) \sum_{m|(n/d)} \mu(m) \qquad (1.29)$$

when we interchange orders of summation. The inner sum is zero by eqn (1.27), unless $d = n$, when only one term $f(d)$ remains. The converse is proved similarly.

We can also define an *additive function* to be an arithmetical function $f(m)$ with

$$f(mn) = f(m)+f(n) \quad \text{when } (m,n) = 1. \qquad (1.30)$$

The simplest examples are $\log m$ and the number of prime factors of m.

There are useful arithmetical functions that are neither multiplicative nor additive. We shall make much use of $\Lambda(m)$, given by

$$\Lambda(m) = \begin{cases} \log p & \text{if } m \text{ is a prime power } p^a,\ a \geqslant 1, \\ 0 & \text{if } m \text{ is not a prime power.} \end{cases} \qquad (1.31)$$

It satisfies the equation

$$\sum_{d|m} \Lambda(d) = \log m. \qquad (1.32)$$

We could have used eqn (1.32) to define $\Lambda(m)$ and recovered the definition (1.31) by Möbius's inversion formula (1.28).

2

SOME SUM FUNCTIONS

THE study of the sum functions of arithmetical functions is important in analytic number theory. For instance, we shall treat many of the properties of prime numbers by using the sum function

$$\psi(x) = \sum_{m \leqslant x} \Lambda(m). \tag{2.1}$$

Our object is to express the sum function as a smooth main term (a power of x or of $\log x$, for example) plus an error term. In place of the cumbersome

$$|f(x)| = O(g(x)), \tag{2.2}$$

we shall often write

$$f(x) \ll g(x), \tag{2.3}$$

and other asymptotic inequalities similarly. Some sum functions can be estimated by writing the arithmetical function as a sum over divisors and rearranging. In this chapter we shall give examples of this method.

From the theory of the logarithmic function we borrow the relation

$$\sum_{m=1}^{M} \left(\frac{1}{m} - \log\left(\frac{m+1}{m}\right)\right) = \gamma + O\left(\frac{1}{M}\right), \tag{2.4}$$

where γ is a constant lying between $\frac{1}{2}$ and 1. We deduce the useful formula

$$\sum_{m=1}^{M} \frac{1}{m} = \log(M+1) + \gamma + O(M^{-1}). \tag{2.5}$$

Our first example is an asymptotic formula for

$$\Phi(x) = \sum_{m \leqslant x} \varphi(m) \tag{2.6}$$

as x tends to infinity. Since $\varphi(m)$ is the number of integers r with $1 \leqslant r \leqslant m$ and $(r, m) = 1$, eqn (1.27) gives

$$\sum_{\substack{d|m \\ d|r}} \mu(d) = \begin{cases} 1 & \text{if } (r, m) = 1, \\ 0 & \text{if not.} \end{cases} \tag{2.7}$$

Hence
$$\varphi(m) = \sum_{r=1}^{m} \sum_{\substack{d|m \\ d|r}} \mu(d), \tag{2.8}$$

and so
$$\begin{aligned}
\Phi(x) &= \sum_{m\leqslant x} \sum_{r=1}^{m} \sum_{\substack{d|m \\ d|r}} \mu(d) \\
&= \sum_{d\leqslant x} \mu(d) \sum_{\substack{m\leqslant x \\ m\equiv 0 (\mathrm{mod}\, d)}} \sum_{\substack{r=1 \\ r\equiv 0 (\mathrm{mod}\, d)}}^{m} 1 \\
&= \sum_{d\leqslant x} \mu(d) \sum_{\substack{m\leqslant x \\ m\equiv 0 (\mathrm{mod}\, d)}} m/d \\
&= \sum_{d\leqslant x} \mu(d) \sum_{n\leqslant x/d} n,
\end{aligned} \tag{2.9}$$

where we have written n for m/d, and the sum over n is
$$\tfrac{1}{2}(x/d)^2 + O(x/d). \tag{2.10}$$

We now write eqn (2.9) as
$$\begin{aligned}
\Phi(x) &= \sum_{d\leqslant x} \mu(d)\left(\frac{1}{2}\frac{x^2}{d^2} + O\left(\frac{x}{d}\right)\right) \\
&= \tfrac{1}{2}x^2 \sum_{d\leqslant x} \frac{\mu(d)}{d^2} + O\left(x \sum_{d\leqslant x} \frac{1}{d}\right) \\
&= \tfrac{1}{2}x^2 \sum_{1}^{\infty} \frac{\mu(d)}{d^2} + O(x) + O(x \log x) \\
&= cx^2 + O(x \log x),
\end{aligned} \tag{2.11}$$

where the constant c can be evaluated; in the notation of Chapter 5 it is $\{2\zeta(2)\}^{-1}$, which can be shown to be $3\pi^{-2}$. The size of the error term is highly satisfactory: when x is a prime, $\Phi(x)$ has a discontinuity $x-1$, and the actual error in (2.11) is $\gg x$ infinitely often. Even so, the upper bound $O(x \log x)$ for the error term in (2.11) has been improved a little.

When we first apply the same method to
$$D(x) = \sum_{m\leqslant x} d(m), \tag{2.12}$$

we find that
$$\begin{aligned}
D(x) &= \sum_{m\leqslant x} \sum_{r|m} 1 \\
&= \sum_{r\leqslant x} \left(\frac{x}{r} + O(1)\right) \\
&= x(\log x + \gamma + O(x^{-1})) + O(x) \\
&= x \log x + O(x),
\end{aligned} \tag{2.13}$$

by (2.5). The error term here is much larger in proportion to the main term than in (2.12); in fact a little cunning enables us to improve (2.13). We let y be the positive integer for which $y^2 \leqslant x \leqslant (y+1)^2$, and write $m = qr$ in eqn (2.12), so that

$$\begin{aligned} D(x) &= \sum_{qr \leqslant x} 1 \\ &= \sum_{q \leqslant y} \sum_{r \leqslant x/q} 1 + \sum_{r \leqslant y} \sum_{q \leqslant x/r} 1 - \sum_{q \leqslant y} \sum_{r \leqslant y} 1 \\ &= 2 \sum_{q \leqslant y} (x/q + O(1)) - y^2 \\ &= x \log x + (2\gamma - 1)x + O(x^{\frac{1}{2}}), \end{aligned} \tag{2.14}$$

when we substitute (2.5) and the value of y.

Dirichlet's divisor problem is to improve the exponent of x in the error term of (2.14). I. M. Vinogradov (1955) has shown that the error term in (2.14) is $\ll x^{\frac{1}{3}} \log^2 x$ by elementary arguments. Van der Corput and others have obtained estimates $\ll x^{\delta}$ with values of δ a little less than $\frac{1}{3}$; their method involves writing the error in (2.14) as a contour integral. On the other side, it has been shown that the error term is $\gg x^{\delta}$ for each $\delta < \frac{1}{4}$. The limitation on the accuracy of (2.14) is not so easily explained as that of (2.11): it is not difficult to show that

$$d(m) \ll m^t \tag{2.15}$$

for each $t > 0$, and so the error in (2.14) must often be very much greater than any individual step in the value of $D(x)$.

To illustrate more difficult examples, we consider

$$\sum_{m \leqslant x} d^2(m)/m. \tag{2.16}$$

Let $d_r(m)$ be the number of ways of writing the positive integer m as a product of r positive integers (so that $d(m) = d_2(m)$). By induction on (2.14),

$$\begin{aligned} D_r(x) &= \sum_{m \leqslant x} d_r(m) \\ &= \sum_{uv \leqslant x} d_{r-1}(v) \\ &= \frac{x \log^{r-1} x}{(r-1)!} + O(x \log^{r-2} x), \end{aligned} \tag{2.17}$$

and by partial summation

$$\begin{aligned} \sum_{m \leqslant x} \frac{d_4(m)}{m} &= \sum_{m \leqslant x} \frac{D_4(x)}{m(m+1)} + O\left(\frac{\log^3 x}{x}\right) \\ &= \tfrac{1}{24} \log^4 x + O(\log^3 x). \end{aligned} \tag{2.18}$$

We have chosen to compare $d^2(m)$ with $d_4(m)$ since, when m is a prime power p^a,
$$d_4(m) = \tfrac{1}{6}(a+1)(a+2)(a+3), \tag{2.19}$$
which is equal to $d^2(m)$ if $a = 0$ or 1. The next step is to find a function $b(m)$ for which
$$d^2(m) = \sum_{uv=m} d_4(u)b(v). \tag{2.20}$$
Since
$$\tfrac{1}{6}(a+1)(a+2)(a+3)-\tfrac{1}{6}(a-1)a(a+1) = (a+1)^2, \tag{2.21}$$
the choice $b(1) = 1$, $b(p^2) = -1$, $b(p^a) = 0$, for prime powers p^a with a not zero or two, satisfies eqn (2.20) when m is a prime power. If we complete the definition of $b(m)$ by making it multiplicative, then (2.20) holds for all m. The choice is thus
$$b(m) = \begin{cases} \mu(n) & \text{if } m = n^2, \\ 0 & \text{if } m \text{ is not a perfect square.} \end{cases} \tag{2.22}$$

We now complete the proof. Equations (2.20) and (2.22) give
$$\sum_{m\leqslant x} \frac{d^2(m)}{m} = \sum_{uv\leqslant x} \frac{d_4(u)b(v)}{uv}$$
$$= \sum_{t^2\leqslant x} \frac{\mu(t)}{t^2} \sum_{u\leqslant x/t^2} \frac{d_4(u)}{u}. \tag{2.23}$$
When we substitute (2.18) and the value $6\pi^{-2}$ of $\sum_1^\infty \mu(t)t^{-2}$ into this, we have
$$\sum_{m\leqslant x} \frac{d^2(m)}{m} = \left(\frac{1}{4\pi^2}+o(1)\right)\log^4 x. \tag{2.24}$$
We require (2.24) and upper estimates for similar sums in the later work. Any $\ll$ estimate for a sum involving divisor functions that we quote will be a corollary of (2.14) or of (2.24), possibly using partial summation.

The method we employ in this chapter can be summarized as follows. To work out a general sum function
$$F(x) = \sum_{m\leqslant x} f(m), \tag{2.25}$$
we try to write
$$f(m) = \sum_{uv=m} a(u)b(v), \tag{2.26}$$
where we have an asymptotic formula for the sum function of the $a(m)$, and $b(m)$ is in some sense smaller than $a(m)$. A necessary condition is that
$$b(p) = o(|a(p)|). \tag{2.27}$$
Our determination of $\Phi(x)$ depended on the equation
$$\varphi(m) = \sum_{uv=m} u\mu(v), \tag{2.28}$$
which is of the form (2.26); and eqn (2.20) is clearly of this form.

3

CHARACTERS

THE reduced residue classes mod q form a group under multiplication, and the characters mod q correspond to the maps $m \to \chi(m)$ from this group to the group of complex numbers of unit modulus under multiplication. We define a group operation on the set of the $\varphi(q)$ characters mod q: the product $\chi_1\chi_2$ of two characters χ_1 and χ_2 mod q is the map $m \to \chi_1(m)\chi_2(m)$. The unit of this group is the trivial character mod q. When the group of reduced residues and the group of characters are each expressed as a direct sum of cyclic groups, we can see that they are isomorphic and that the homomorphisms of the group of characters to the complex numbers are given by $\chi \to \chi(m)$, where m runs through the reduced residues mod q. Two finite Abelian groups related in this way are said to be *dual*.

We have already seen that

$$\sum_{m \bmod q} e_q(am) = \begin{cases} q & \text{if } a \equiv 0 \pmod{q}, \\ 0 & \text{if } a \not\equiv 0 \pmod{q}, \end{cases} \tag{3.1}$$

for the maps $m \to e_q(am)$ from the group of residue classes $m \pmod{q}$ under addition to the complex numbers of unit modulus. We have similar results for the characters $\chi(m)$ mod q:

$$\sum_{m \bmod q} \chi(m) = \begin{cases} \varphi(q) & \text{if } \chi \text{ is trivial}, \\ 0 & \text{if } \chi \text{ is non-trivial}. \end{cases} \tag{3.2}$$

Here $\chi(m)$ is non-zero when the residue class $m \pmod{q}$ is reduced, and thus when m is a member of the group of reduced residue classes under multiplication. Now eqn (3.1) has a dual interpretation, in terms of residue classes $a \pmod{q}$ and maps $a \to e_q(am)$. This corresponds to

$$\sum_{\chi \bmod q} \chi(m) = \begin{cases} \varphi(q) & \text{if } m \equiv 1 \pmod{q}, \\ 0 & \text{if } m \not\equiv 1 \pmod{q}. \end{cases} \tag{3.3}$$

The proof of eqn (3.2) runs as follows. If χ is trivial, then $\chi(m) = 1$ when $(m, q) = 1$ and 0 otherwise, so that the sum on the left-hand side of (3.2) is the definition (1.10) of $\varphi(q)$. If χ is non-trivial, there is an integer r for which $(r, q) = 1$ but $\chi(r)$ is not unity. We multiply the sum on the left of eqn (3.2) by $\chi(r)$; this gives $\sum \chi(mr)$, where m runs through a complete set of residues mod q. Since $(r, q) = 1$, mr also runs through a complete set of residues mod q, and the sum is unchanged. This contradicts the choice of r with $\chi(r) \neq 1$ if the sum is not zero. Similarly, in eqn (3.3) if m is not congruent to unity mod q there is a character χ_1 with $\chi_1(m) \neq 1$. When we multiply the left-hand side of (3.3) by $\chi_1(m)$, $\sum \chi(m)\chi_1(m)$ is a sum over all characters mod q, and again this sum must be zero, for we have multiplied by a constant that is not unity but have succeeded only in permuting the terms of the sum. Of course, eqns (3.1), (3.2), and (3.3) are essentially special cases of a theorem on dual finite Abelian groups.

An important notion is the *propriety* of characters. Let q_2 be a multiple of q_1, and χ_1 a character mod q_1. The group of reduced residue classes mod q_2 maps homomorphically onto the corresponding group mod q_1, and we define a character χ_2 mod q_2 by the equation

$$\chi_2(m) = \begin{cases} \chi_1(m) & \text{if } (m, q_2) = 1, \\ 0 & \text{if } (m, q_2) > 1. \end{cases} \tag{3.4}$$

We note that χ_1 and χ_2 are different arithmetical functions. If $q_1 = 3$, $q_2 = 6$, and χ_1 takes the values

$$1,\ -1,\ 0,\ 1,\ -1,\ 0 \tag{3.5}$$

for $m = 1, 2, 3, 4, 5, 6$, then χ_2 takes the values

$$1,\ 0,\ 0,\ 0,\ -1,\ 0, \tag{3.6}$$

since χ_2 is zero when m is a multiple of 2 as well as when $m \equiv 0 \pmod 3$. When χ_2 is constructed by eqn (3.4), we say that χ_1 mod q_1 *induces* χ_2 mod q_2, and, if $q_2 \neq q_1$, that χ_2 mod q_2 is *improper*. A *proper character* mod q (also called a *primitive character*) is one that is not induced by a character mod d for any divisor d of q other than q itself. The smallest f for which a character χ_1 mod f induces χ mod q is called the *conductor* of χ. The customary letter f is the initial of a German word for a tram conductor.

We shall now discuss *Gauss's sum* $\tau(\chi)$, defined by

$$\tau(\chi) = \sum_{m \bmod q} \chi(m) e_q(m). \tag{3.7}$$

In this curious expression, the factors correspond to the multiplicative group of reduced residues mod q and the additive group of all residues mod q. The absolute value of $\tau(\chi)$ is found below. Gauss (preceding Dirichlet) considered only characters χ for which $\chi(m)$ takes the values ± 1 and zero only. In this case τ^2 is real, but it is still not easy to find the argument of $\tau(\chi)$.

We shall use $\tau(\chi)$ to remove characters from a summation. With $\bar{\chi}(m)$ denoting the complex conjugate of $\chi(m)$ (the inverse of χ in the group of characters), eqn (3.7) gives

$$\tau(\bar{\chi})\chi(m) = \sum_{a \bmod q} \bar{\chi}(a)e_q(am) \tag{3.8}$$

whenever $(m, q) = 1$. If $\tau(\bar{\chi})$ is non-zero, we can use (3.8) to change a summation over $\chi(m)$ to a summation over the exponential maps $e_q(am)$, which are easier to manipulate. A defect in eqn (3.8) is the condition $(m, q) = 1$. If, however, χ is proper mod q eqn (3.8) holds for all integers m. We must show that the sum on the right-hand side of (3.8) is zero when

$$m = tn, \qquad q = tr, \qquad t > 1. \tag{3.9}$$

In this case,

$$\sum_{a \bmod q} \bar{\chi}(a)e_q(am) = \sum_{a \bmod q} \bar{\chi}(a)e_r(an). \tag{3.10}$$

Since χ is not induced by any character mod r, there is an integer b with $(b, q) = 1$, $b \equiv 1 \pmod{r}$, but $\chi(b) \neq 1$. Our standard proof now applies. Multiplication by $\bar{\chi}(b)$ permutes the residue classes in the sum on the left of eqn (3.10), but multiplies the value of the sum by a constant that is not unity. The sums in (3.10) and (3.8) are thus zero if χ is proper mod q and $(m, q) > 1$.

For eqn (3.8) to be of use we must be sure that $\tau(\chi)$ is non-zero. When χ is proper mod q there is an elegant demonstration. For each m mod q,

$$|\chi(m)\tau(\bar{\chi})|^2 = \sum_{a \bmod q} \sum_{b \bmod q} \bar{\chi}(a)\chi(b)e_q(am-bm). \tag{3.11}$$

Hence

$$\sum_{m \bmod q} |\chi(m)|^2 |\tau(\bar{\chi})|^2 = \sum_{a \bmod q} \sum_{b \bmod q} \bar{\chi}(a)\chi(b) \sum_{m \bmod q} e_q(am-bm). \tag{3.12}$$

The inner sum is zero by eqn (3.1), unless $a \equiv b$, when it is q, so that

$$|\tau(\bar{\chi})|^2 \sum_{m \bmod q} |\chi(m)|^2 = q \sum_{a \bmod q} |\chi(a)|^2. \tag{3.13}$$

Since $\bar{\chi}$ is proper mod q if and only if its inverse χ is proper mod q, we deduce that

$$|\tau(\chi)|^2 = q \quad \text{for } \chi \text{ proper mod } q. \tag{3.14}$$

Equation (3.8) thus applies when χ is proper mod q, the factor $\tau(\bar{\chi})$ being non-zero.

We conclude with the case when χ mod q has conductor f, and $q = fg$ where $g > 1$. If the lowest common multiple $h = [f, g]$ of f and g is not q itself, χ is not merely induced by some character χ_2 mod h, but is actually equal to χ_2, since any integer prime to h is already prime to $fg = q$. Replacement of m in the sum in eqn (3.7) by $m+h$ permutes the residue classes mod q, but multiplies $\tau(\chi)$ by $e_q(h)$, which is not unity. In this case therefore $\tau(\chi)$ is zero.

When $(f,g)=1$ we invoke the Chinese remainder theorem. By eqn (1.2) there are integers u, v with

$$fu+gv=1. \tag{3.15}$$

Residue classes a (mod fg) correspond to pairs of classes b (mod f), c (mod g) according to the relation

$$a\equiv cfu+bgv \pmod{fg}. \tag{3.16}$$

In (3.16), a (mod fg) is a reduced class if and only if both b (mod f) and c (mod g) are reduced, and thus

$$\begin{aligned}\tau(\chi) &= \sum_{a \bmod q}^{*} \chi(a)e_q(a)\\ &= \sum_{b \bmod f}^{*}\sum_{c \bmod g}^{*} \chi(cfu+bgv)e_g(cu)e_f(bv)\\ &= \sum_{b \bmod f}^{*} \chi(b)e_f(bv)\sum_{c \bmod g}^{*} e_g(cu)\\ &= \bar{\chi}_1(v)\tau(\chi_1)c_g(u),\end{aligned} \tag{3.17}$$

where χ_1 mod f is the character inducing χ mod q, and $c_g(u)$ is Ramanujan's sum, defined in eqn (1.11).

We now proceed to compute Ramanujan's sum. By eqns (1.11) and (1.27),

$$\begin{aligned}c_g(u) &= \sum_{a \bmod g} e_g(au)\sum_{\substack{d\mid a\\ d\mid g}}\mu(d)\\ &= \sum_{d\mid g}\mu(d)\sum_{b \bmod g/d} e_{g/d}(bu),\end{aligned} \tag{3.18}$$

where we have written $a=bd$. From eqn (3.1) the inner sum is zero unless u is a multiple of g/d. Writing $h=g/d$, we have

$$c_g(u)=\sum_{\substack{h\mid g\\ h\mid u}} h\mu(g/h). \tag{3.19}$$

It is possible to continue and to express $c_g(u)$ in terms of Euler's φ function. In our application, eqn (3.15) ensures that $(g,u)=1$ and thus that $c_g(u)$ is $\mu(g)$, which is itself zero if g has a repeated prime factor. Ramanujan's sum with $(g,u)=1$ can be regarded as Gauss's sum for the trivial character χ_0 mod g, for which $\chi_0(m)=1$ whenever $(g,m)=1$.

In this chapter, we have shown that $\tau(\chi)$ is zero unless $g=q/f$ is composed solely of those primes whose squares do not divide q, in which case

$$|\tau(\chi)|^2=f. \tag{3.20}$$

4

PÓLYA'S THEOREM

'About as big as Piglet,' he said to himself sadly. 'My favourite size. Well, well.'

I. 85

FROM eqn (3.2) we see that the sum function

$$X(x) = \sum_{m \leqslant x} \chi(m) \tag{4.1}$$

is bounded when χ is a non-trivial character mod q. Since the sum over any q consecutive integers is zero, the absolute value of $X(x)$ can be at most $\frac{1}{2}q$. Pólya (1918) proved the following sharper result.

THEOREM. *Let χ be a non-trivial character* mod q *with conductor f. Then*

$$\left|\sum_{m \leqslant x} \chi(m)\right| \leqslant (\pi^{-1}+o(1))f^{\frac{1}{2}}d(q/f)\log f, \tag{4.2}$$

where the term $o(1)$ is to be interpreted as $f \to \infty$.

Pólya's theorem was discovered independently by I. M. Vinogradov (1955, chapter 3, example 12), with a different constant in the upper bound. Later proofs have been given by Linnik and Rényi (1947) and by Knapowski in an unpublished manuscript. We shall follow Pólya's argument, as it is the most precise and can easily be adapted to show that the sum in (4.2) is frequently $\gg f^{\frac{1}{2}}$.

First we introduce some notation. If α is a real number, we write $[\alpha]$ for the largest integer not exceeding α, and $\|\alpha\|$ for the distance from α to the nearest integer, so that

$$[\alpha] = \max_{m \leqslant \alpha} m, \tag{4.3}$$

$$\|\alpha\| = \min |m-\alpha|, \tag{4.4}$$

where the maximum in (3.3) and the minimum in (3.4) are over all integers m. We now state a lemma.

LEMMA. *The Fourier series*

$$H(\alpha) = \sum_{\substack{m=-\infty \\ m \neq 0}}^{\infty} \frac{e(m\alpha)}{-2\pi i m} \tag{4.5}$$

converges to $$\begin{cases} \alpha-[\alpha]-\frac{1}{2} & \textit{if } \alpha \textit{ is not an integer,} \\ 0 & \textit{if } \alpha \textit{ is an integer,} \end{cases} \tag{4.6}$$

and the partial sums satisfy the relation

$$\left| H(\alpha)+\sum_{\substack{-M \\ m\neq 0}}^{M} \frac{e(m\alpha)}{2\pi i m} \right| \leqslant \frac{1}{2\pi M\|\alpha\|} \tag{4.7}$$

when α *is not an integer.*

Proof. We prove (4.7). Since $H(\alpha)$ has period 1 and $H(-\alpha) = -H(\alpha)$, we suppose that $0 < \alpha \leqslant \frac{1}{2}$. Now

$$(2\pi i m)^{-1}e(m\alpha)-(2\pi i m)^{-1}(-1)^m = \int_{\frac{1}{2}}^{\alpha} e(mt)\,dt, \tag{4.8}$$

and thus $$\sum_{\substack{-M \\ m\neq 0}}^{M} (2\pi i m)^{-1}e(m\alpha)+\tfrac{1}{2}-\alpha = \int_{\frac{1}{2}}^{\alpha} \sum_{-M}^{M} e(mt)\,dt$$

$$= \int_{\frac{1}{2}}^{\alpha} \frac{e(Mt+\frac{1}{2}t)-e(-Mt-\frac{1}{2}t)}{e(\frac{1}{2}t)-e(-\frac{1}{2}t)}\,dt$$

$$= \int_{\frac{1}{2}}^{\alpha} \frac{\sin(2M+1)\pi t}{\sin \pi t}\,dt, \tag{4.9}$$

which by one of the mean-value theorems for integrals does not exceed in modulus

$$\left| \int_{\frac{1}{2}}^{\beta} \frac{\sin(2M+1)\pi t}{\sin \pi\alpha}\,dt \right| \leqslant \frac{1}{2\alpha}\frac{2}{(2M+1)\pi} \leqslant \frac{1}{2\pi M\alpha} \tag{4.10}$$

for some β between α and $\frac{1}{2}$. This has proved (4.7), which gives (4.6) if α is not an integer; if α is an integer every term in $H(\alpha)$ is zero.

We now return to the sum in eqn (4.1). Since χ is non-trivial, we have

$$\sum_{rq+1}^{rq+q} \chi(m) = 0, \tag{4.11}$$

and so we can replace the sum $X(x)$ by a sum of the same form with $0 \leqslant x < q$. If now $x > \frac{1}{2}q$, we can replace m by $-q+m$ and obtain a sum with $0 \leqslant x \leqslant \frac{1}{2}q$, multiplied by $-\chi(-1)$. Thus we can suppose that $0 \leqslant x \leqslant \frac{1}{2}q$ in (4.1). We now use $H(\alpha)$ to construct $G(\alpha)$, where

$$G(\alpha) = \begin{cases} 1 & \text{if } 0 < \alpha < x/q, \\ \frac{1}{2} & \text{if } \alpha = 0 \text{ or } \alpha = x/q, \\ 0 & \text{if } x/q < \alpha < 1, \end{cases} \tag{4.12}$$

using the equation

$$G(\alpha) = x/q + H(\alpha - x/q) - H(\alpha). \tag{4.13}$$

We now have

$$\chi(1)+\chi(2)+\ldots+\tfrac{1}{2}\chi(x) = \sum_{m=1}^{q} \chi(m)G(m/q). \tag{4.14}$$

When we use eqn (4.7) to truncate the sums for $H(\alpha - x/q)$ and $H(\alpha)$ in (4.14), the total error in modulus is at most

$$\begin{aligned} 2\sum_{m=1}^{q} (2\pi M\|m/q\|)^{-1} &\leqslant 4\sum_{m=1}^{\frac{1}{2}q} \frac{q}{2\pi m M} \\ &\leqslant 2\pi^{-1}M^{-1}q\log q. \end{aligned} \tag{4.15}$$

We now have finite sums to manipulate. Writing

$$G(\alpha) = \sum_{-\infty}^{\infty} a(m)e(m\alpha), \tag{4.16}$$

we have to consider

$$\sum_{r=1}^{q} \chi(r) \sum_{-M}^{M} e_q(mr). \tag{4.17}$$

Supposing first that χ is proper mod q, we have from eqn (3.8)

$$\sum_{-M}^{M} a(m) \sum_{r=1}^{q} \chi(r)e_q(mr) = \sum_{-M}^{M} a(m)\bar{\chi}(m)\tau(\chi). \tag{4.18}$$

Since χ is non-trivial $a(0)\chi(0)$ is zero, and by eqns (4.5) and (4.13), for $m \neq 0$,

$$|a(m)| \leqslant |\pi m|^{-1}. \tag{4.19}$$

The modulus of the expression in (4.18) is now

$$\begin{aligned} &\leqslant q^{\frac{1}{2}}\, 2\sum_{m=1}^{M} (\pi m)^{-1} \\ &\leqslant 2\pi^{-1}q^{\frac{1}{2}}(\log M + O(1)). \end{aligned} \tag{4.20}$$

We choose

$$M = q^{\frac{1}{2}+\delta}, \tag{4.21}$$

so that (4.20) is

$$\pi^{-1}q^{\frac{1}{2}}\log q(1+O(\delta)), \tag{4.22}$$

and after (4.15) the tails of the series give

$$\leqslant (\pi M)^{-1}q\log q = o(q^{\frac{1}{2}}\log q). \tag{4.23}$$

Finally the omitted term $\frac{1}{2}\chi(x)$ is $O(1)$, and we have Pólya's theorem when χ is proper mod q.

If χ is induced by χ_1 proper mod f, where $f < q$, we could complete the proof similarly, but it is easier to deduce this case.

$$\sum_{m \leqslant x} \chi(m) = \sum_{m \leqslant x} \chi_1(m) \sum_{\substack{d|q \\ d|m \\ (d,f)=1}} \mu(d)$$

$$= \sum_{\substack{d|q \\ (d,f)=1}} \mu(d)\chi_1(d) \sum_{m \leqslant x/d} \chi_1(m)$$

$$\leqslant \sum_{\substack{d|q \\ (d,f)=1}} f^{\frac{1}{2}} \log f\{\pi^{-1}+o(1)\}, \tag{4.24}$$

since χ_1 is proper mod f. The number of terms in the sum is at most $d(q/f)$, and we have completed the proof of Pólya's theorem.

As a simple corollary we prove that for each prime $p > 2$ there is an integer m with

$$m \ll p^{\frac{1}{2}} \log p \tag{4.25}$$

for which the congruence

$$u^2 \equiv m \pmod{p} \tag{4.26}$$

has no solution. Since 1, 4, 9,..., $\frac{1}{4}(p-1)^2$ are distinct mod p, there are $\frac{1}{2}(p-1)$ reduced residue classes that are congruent to squares mod p, and these form a subgroup of index 2. There is therefore a character χ mod p with $\chi(m) = 1$ when m is congruent to the square of a reduced residue and -1 when $(m, q) = 1$ but m is not congruent to a square. We choose

$$x > p^{\frac{1}{2}} \log p \tag{4.27}$$

in (4.2). Not all terms in the sum (4.1) can be non-negative if p is sufficiently large, and so there is an $m \leqslant p^{\frac{1}{2}} \log p$ with $\chi(m) = -1$. The exponent $\frac{1}{2}$ in (4.25) can be improved, but the conjecture that the asymptotic inequality

$$m \ll p^{\delta} \tag{4.28}$$

for each $\delta > 0$ holds in place of (4.25) has not yet been proved or confounded.

5

DIRICHLET SERIES

'Well,' said Owl, 'the customary procedure in such cases is as follows.'

'What does Crustimoney Proseedcake mean?' said Pooh. 'For I am a Bear of Very Little Brain, and long words bother me.'

'It means the Thing to Do.'

I. 48

A *Dirichlet series* is an analytic function of the complex variable $s = \sigma + it$ defined by a series

$$f(s) = \sum_{m=1}^{\infty} a(m)m^{-s}, \tag{5.1}$$

or a generalization thereof. All the Dirichlet series that we need are special cases of (5.1). If eqn (5.1) converges at $s_0 = \sigma_0 + it_0$, then $|a(m)m^{-s_0}|$ is bounded. This simple observation is the basis for the theory of convergence of Dirichlet series. By partial summation (5.1) converges whenever $\sigma > \sigma_0$ and converges absolutely when $\sigma > \sigma_0 + 1$. The convergence is uniform in a half-plane $\sigma \geqslant \alpha$, provided $\alpha > \sigma_0 + 1$. We see that the region of definition of $f(s)$ is a half-plane bounded to the left by some vertical line; this line is called the *abscissa of convergence.*

If
$$A(x) = \sum_{m \leqslant x} a(m) \tag{5.2}$$

is the sum function of the coefficients in (5.1), then

$$f(s) = \int_1^{\infty} s x^{-s-1} A(x)\, dx. \tag{5.3}$$

Formula (5.3) can be inverted: from $f(s)$ we can recover the sum function $A(x)$ of the coefficients. Let α and u be positive real numbers. Then

$$\frac{1}{2\pi i} \int_{\alpha - i\infty}^{\alpha + i\infty} \frac{u^s\, ds}{s} = \begin{cases} 0 & \text{if } 0 < u < 1, \\ \frac{1}{2} & \text{if } u = 1, \\ 1 & \text{if } u > 1. \end{cases} \tag{5.4}$$

For $u \neq 1$ we consider the integral from $\alpha - iT_1$ to $\alpha + iT_2$. When $u > 1$, this is equal to the integral round the three remaining sides of the rectangle whose other corners are $R/\log u+iT_2$, $R/\log u-iT_1$. The modulus of the integral in eqn (5.4) is thus

$$\leqslant \frac{R^{-1}e^{-R}}{2\pi}+\frac{u^{-\alpha}-e^{-R}}{2\pi T_1 \log u}+\frac{u^{-\alpha}-e^{-R}}{2\pi T_2 \log u}, \tag{5.5}$$

a number which tends to zero as R, T_1, and T_2 tend to $+\infty$. When $u < 1$, $R/\log u$ is negative, and we must add the residue from the pole of s^{-1} at $s = 0$; this gives unity. Finally when $u = 1$ we define the value of the integral (5.4) to be the limit of the integral from $\alpha-iT$ to $\alpha+iT$ when $T \to \infty$. This reduces to an inverse tangent integral.

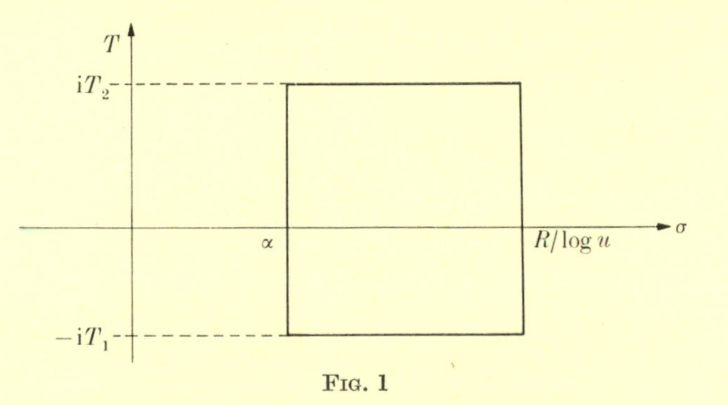

FIG. 1

If $f(s)$ defined by eqn (5.1) converges uniformly in t on the line $\sigma = \alpha$, then for $x > 0$ term-by-term integration gives

$$\frac{1}{2\pi i}\int\limits_{\alpha-i\infty}^{\alpha+i\infty} x^s s^{-1} f(s)\,ds = \sum_{m<x} a(m)+\tfrac{1}{2}a(x), \tag{5.6}$$

where the last term occurs only if x is an integer. There are many integral transforms from Dirichlet series to their coefficient sums, all proved by the same method. The simplest one after (5.6) itself is

$$\frac{1}{2\pi i}\int\limits_{\alpha-i\infty}^{\alpha+i\infty} \frac{x^s f(s)}{s(s+1)}\,ds = \sum_{m<x}\left(1-\frac{m}{x}\right)a(m). \tag{5.7}$$

A special class of Dirichlet series occurs naturally in number theory. The functions are known as *L-functions* after Dirichlet's functions

$$L(s,\chi) = \sum_{m=1}^{\infty} \chi(m)m^{-s}, \tag{5.8}$$

where χ is a Dirichlet's character to some modulus q, or as *zeta functions* after Riemann's function

$$\zeta(s) = \sum_{m=1}^{\infty} m^{-s}, \tag{5.9}$$

which is the special case of Dirichlet's definition when $q = 1$ and χ is trivial.

L-functions are defined by two properties. First, the coefficients $a(m)$ are multiplicative, so that Euler's product identity

$$\sum_{m=1}^{\infty} \frac{a(m)}{m^s} = \prod_p \left(1+\frac{a(p)}{p^s}+\frac{a(p^2)}{p^{2s}}+\dots\right) \tag{5.10}$$

holds in a half-plane $\sigma \geqslant \alpha$ in which one side of eqn (5.10) converges absolutely. If the product in (5.10) converges, $f(s)$ can be zero only when one of the factors on the right-hand side of (5.10) is zero. The convergence of the left-hand side of (5.10) alone does not imply that of the product; $L(s, \chi)$ with χ non-trivial has a series (5.8) converging for $\sigma > 0$, but the function itself has zeros in $\sigma \geqslant \frac{1}{2}$, preventing the product from converging in $0 < \sigma < \frac{1}{2}$.

The second defining property is that $f(s)$ should have a functional equation

$$f(s)G(s) = f^*(r-s)G^*(r-s), \tag{5.11}$$

where r is a positive integer, $G(s)$ is essentially a product of gamma functions, and the operation * has $(f^*)^* = f$ and $(G^*)^* = G$. As an example, in the functional equation for $L(s, \chi)$ in Chapter 11, $L^*(s, \chi)$ is $L(s, \bar{\chi})$. An important conjecture about L-functions is the *Riemann hypothesis* that if $f(s)$ satisfies eqns (5.10) and (5.11) then all zeros of $f(s)G(s)$ have real part $\frac{1}{2}r$. The truth or falsity of this hypothesis is not settled for any L-function.

Two generalizations that are often called zeta functions are

$$\sum_{m=1}^{\infty} (m+\delta)^{-s}, \tag{5.12}$$

where δ is a fixed real number, and

$$\sum_{m=1}^{\infty} r(m)m^{-s}, \tag{5.13}$$

where $r(m)$ is the number of representations of m by a positive definite quadratic form. Except in special cases these fail to have a product formula of the form (5.10), and not all of their zeros lie on the appropriate line. Some authors even use 'zeta function' as a synonym for 'Dirichlet series'.

In Chapter 11 we shall obtain analytic continuations of $\zeta(s)$ and other L-functions over the whole plane. Since the sum function $X(x)$ formed with a non-trivial character χ is bounded, by partial summation (5.8) converges for $\sigma > 0$ except when χ is trivial. Similarly, the function

$$\sum_{m=1}^{\infty} (-1)^{m-1} m^{-s} = (1-2^{1-s})\zeta(s) \tag{5.14}$$

converges for $\sigma > 0$ and provides an analytic continuation for $\zeta(s)$. When we make $s \to 1$ in (5.14), we see that $\zeta(s)$ has a pole of residue 1 at $s = 1$. When we put $f(s) = \zeta(s)$ in (5.6), the integrand has a simple pole at $s = 1$ with residue x. The value of the right-hand side of eqn (5.6) is between $x-1$ and x. If we deform the contour in (5.6) so that it passes to the left of the pole, the residue makes the main contribution, and the contour integral left over is bounded. Let

$$\psi(x) = \sum_{m<x} \Lambda(m) \tag{5.15}$$

and

$$M(x) = \sum_{m<x} \mu(m), \tag{5.16}$$

which are the coefficient sums of $-\zeta'(s)/\zeta(s)$ and of $1/\zeta(s)$ (we shall prove this below). The function $-\zeta'(s)/\zeta(s)$ also has a pole of residue 1 at $s = 1$, but $1/\zeta(s)$ does not. If the corresponding contour integrals were negligible, we should have

$$\psi(x) = x + o(x), \tag{5.17}$$

$$M(x) = o(x). \tag{5.18}$$

These are forms of the prime-number theorem, which we shall prove in Part II.

Writing $m = fg$, we have

$$\sum_{m=1}^{\infty} m^{-s} \sum_{f|m} a(f)b(m/f) = \Big(\sum_{f=1}^{\infty} a(f)f^{-s}\Big)\Big(\sum_{g=1}^{\infty} b(g)g^{-s}\Big). \tag{5.19}$$

If $b(g) = 1$ and $a(f) = \mu(f)$ for each pair of integers f, g,

$$\zeta(s) \sum_{m=1}^{\infty} \mu(m)m^{-s} = \sum_{m=1}^{\infty} m^{-s} \sum_{f|m} \mu(f) = 1 \tag{5.20}$$

from eqn (1.27). Since eqns (5.3) and (5.6) imply that expansions in Dirichlet series are unique, we have shown that the series on the left-hand side of (5.20) represents $1/\zeta(s)$ wherever it converges. Similarly, using eqns (5.19) and (1.32) we can check that $-\zeta'(s)/\zeta(s)$ has a Dirichlet series with coefficients $\Lambda(m)$.

For fifty years (1898–1948) the only proofs known of eqns (5.17) and (5.18) used contour integration and other complex-variable techniques. In 1948, Selberg and Erdös gave a real proof of (5.18) (see Hardy and

Wright (1960), Chapter 22). The real-variable approach is not so well understood, and the strongest forms of (5.17) and (5.18) (those in which the error term is smallest) have been obtained by analytic methods. The form (16.22) in which we shall prove (5.17) is a little stronger than the best so far obtained by Selberg's method.

Apart from the analytic arguments, study of $\log(N!)$ suggests the form (5.17) as a conjecture. By eqn (1.32),

$$\begin{aligned}\log(N!) = \sum_{m\leqslant N}\log m &= \sum_{de\leqslant N}\Lambda(d)\\ &= \sum_{e\leqslant N}\psi(N/e), \qquad (5.21)\end{aligned}$$

where we have written $m = de$ in the first sum. On the other hand, by expression (2.5),

$$\begin{aligned}\sum_{m\leqslant N}\log m &= \sum_{m\leqslant N}\left\{\sum_{r\leqslant m-1}\frac{1}{r}-\gamma+O\left(\frac{1}{m}\right)\right\}\\ &= \sum_{r\leqslant N}\frac{N-r-1}{r}-N\gamma+O(\log N)\\ &= N\log N-N+O(\log N), \qquad (5.22)\end{aligned}$$

which agrees with the result of substituting (5.17) into (5.21). In this way, Gauss was led to conjecture that

$$\sum_{p\leqslant x}1 = \{1+o(1)\}\int_2^x(\log t)^{-1}\,\mathrm{d}t, \qquad (5.23)$$

which is another form of the prime-number theorem. By consideration of the binomial coefficient ${}_{2N}C_N$, it can be shown that $\psi(2N)-\psi(N)$ lies between bounded multiples of N (Hardy and Wright, 1960); but there are too many terms in the sum (5.21) to allow (5.17) to be deduced from (5.22).

6

SCHINZEL'S HYPOTHESIS

And all the good things which an animal likes
Have the wrong sort of swallow or too many spikes.

II. 30

MANY problems in prime-number theory follow a similar pattern. Various constraints are laid on a set of integer unknowns, and we ask whether the integer unknowns can all be prime simultaneously, and whether this happens infinitely often. Many of these problems are subsumed under *Schinzel's hypothesis*: if $f_1(x_1,..., x_n),..., f_m(x_1,..., x_n)$ are polynomials (with integer coefficients) irreducible over the integers, and there is no prime p for which $\prod f_i \equiv 0 \pmod p$ for all sets $x_1,..., x_n$ of residues mod p, then there are infinitely many sets $x_1,..., x_n$ of integers for which the absolute values of $f_1,..., f_m$ are all prime.

There is a conjectured asymptotic formula for the number of sets $x_1,..., x_n$ with $0 \leqslant x_i \leqslant N$ for each i with each of $f_1,..., f_m$ prime, and some bold authors have conjectured that the asymptotic formula can be stated with an error term smaller than the main term by a factor $N^{\frac{1}{2}-\epsilon}$ for each ϵ greater than zero. Thus the simplest case is one polynomial, $f(x) = x$, and the conjecture now states that $\pi(N)$, the number of primes up to N, satisfies the relation

$$\pi(N) = \int_2^N \frac{\mathrm{d}x}{\log x} + O(N^{\frac{1}{2}+\epsilon}), \tag{6.1}$$

a very strong form of the prime-number theorem (5.23). The accuracy of (6.1) seems unattainable. We shall prove later that the hypothesis is true for one linear polynomial $f(x) = qx+a$; this is the prime-number theorem for arithmetical progressions, but the error term in the asymptotic formula will only be shown to be slightly smaller than the leading term.

The next simplest case concerns two linear polynomials, $f_1(x) = x$, $f_2(x) = x-2$. Here the conjectured formula is

$$2\prod_{p\geqslant 3}\{1-(p-1)^{-2}\}\int_2^N (\log t)^{-2}\,\mathrm{d}t + \text{error term}. \tag{6.2}$$

We shall now describe how to write down the conjectured asymptotic formulae. Let

$$S(\alpha) = \sum_{p \leqslant N} e(p\alpha); \tag{6.3}$$

such an expression is called an *exponential sum* or a *trigonometric sum.* By the fundamental relation

$$\int_0^1 e(m\alpha)\,\mathrm{d}\alpha = \begin{cases} 1 & \text{if } m = 0, \\ 0 & \text{if } m \neq 0, \end{cases} \tag{6.4}$$

we see that the number of primes $p \leqslant N$ for which $p-2$ is also prime is

$$\int_0^1 S(\alpha)S(-\alpha)e(-2\alpha)\,\mathrm{d}\alpha. \tag{6.5}$$

We cannot, of course, work out this integral, but we can suggest a plausible value for it. Writing

$$\pi(N; q, b) = \sum_{\substack{p \leqslant N \\ p \equiv b \pmod q}} 1, \tag{6.6}$$

we have

$$S(a/q) = \sum_{b \bmod q} e_q(ab)\pi(N; q, b). \tag{6.7}$$

Now the sum in eqn (6.6) is 0 or 1 if b has a common factor with q. If we make the approximation that the primes are divided equally between the $\varphi(q)$ residue classes b with $(b, q) = 1$, the expression in (6.7) is

$$\{\varphi(q)\}^{-1}\pi(N) \sideset{}{^*}\sum_{b \bmod q} e_q(ab) = \{\varphi(q)\}^{-1}\pi(N)c_q(a), \tag{6.8}$$

where $c_q(a)$ is Ramanujan's sum (1.11). If $(a, q) = 1$ the Ramanujan's sum is just $\mu(q)$, from eqn (3.19). This argument suggests that $S(\alpha)$ has a 'spike' at a/q of height proportional to $\mu(q)/\varphi(q)$, that is, that $|S(\alpha)|$ has a local maximum close to a/q. Now the area under the graph of $|S(\alpha)|^2$ near 0 (which certainly is the site of a spike) may plausibly be written

$$N^{-1}\pi^2(N) + \text{error term}, \tag{6.9}$$

and this can be proved to be true. If we assume further that all the spikes at rational points are the same shape, the spikes at rational points a/q contribute

$$N^{-1}\pi^2(N) \sum_q \mu^2(q)(\varphi(q))^{-2} \sideset{}{^*}\sum_{a \bmod q} e_q(2a) + \text{error term}, \tag{6.10}$$

and the sum over q in (6.10) converges to

$$2 \prod_{p>2} \{1-(p-1)^{-2}\}, \tag{6.11}$$

which gives the main term in eqn (6.2).

For small q the argument above can be made rigorous; but then part of the range of integration in (6.5) does not support spikes. Away from a spike we cannot estimate $S(\alpha)$ except by replacing it by its absolute value; and the spikes with small q contribute very little to $\int_0^1 |S(\alpha)|^2\, d\alpha$. In the integral of $|S(\alpha)|^3$ the spikes do dominate, and by this method I. M. Vinogradov was able to prove that every large odd number is the sum of three primes.

The approach to Schinzel's hypothesis through exponential sums does lead to an upper bound for the number of sets $x_1,..., x_n$ of integers not exceeding N for which $f_1,..., f_m$ are all prime. To explain the method we shall take $n = 1$, so we are considering integers x in the range $1 \leqslant x \leqslant N$ for which $f_1(x),..., f_m(x)$ are all prime. We now work modulo a prime p. Apart from the finite number of x for which one of $f_1(x),..., f_m(x)$ is p, x must be such that none of $f_1(x),..., f_m(x) \equiv 0 \pmod{p}$. This means that x must be confined to certain residue classes mod p. We therefore divide the residue classes mod p into a set $H(p)$ of $f(p)$ forbidden classes and a set $K(p)$ of $g(p) = p-f(p)$ permitted classes; h mod p is forbidden if and only if one of the polynomials $f_i(h)$ is a multiple of p. If x falls into a forbidden class for any prime p smaller than each of the $f_i(x)$, then one at least of the $f_i(x)$ cannot be prime.

The values of x that make $f_1(x),..., f_m(x)$ primes greater than some bound Q form a sifted sequence, in the following sense. The increasing sequence $\mathcal{N}$ of positive integers $n_1, n_2,...$ is *sifted* by the primes $p \leqslant Q$ if for each prime $p \leqslant Q$ there is a set $H(p)$ (possibly empty) of $f(p)$ residue classes mod p into which no member of $\mathcal{N}$ falls. We shall show in Chapter 8 that, if $\mathcal{N}$ satisfies the above condition, the number of members of $\mathcal{N}$ in any interval of N consecutive integers is

$$\frac{N}{\sum_{q \leqslant Q} \mu^2(q) f(q)/g(q)} + \text{error term}, \tag{6.12}$$

where

$$f(q) = q \prod_{p|q} f(p)/p, \tag{6.13}$$

$$g(q) = q \prod_{p|q} \{1-f(p)/p\}. \tag{6.14}$$

We shall work out examples of this upper bound in Chapter 8; in each case the leading term is a multiple of the leading term in the conjectured formula.

Upper bounds of the right order of magnitude were first found by Viggo Brun using combinatorial arguments. Rosser used Brun's method to obtain expression (6.12), which was found in a different way by Selberg. An outline of Selberg's method follows. It rests on the construction of an exponential sum $T(\alpha)$ with the same spikes at rational points as

$$S(\alpha) = \sum_{n_i \leqslant N} e(n_i \alpha) \tag{6.15}$$

is conjectured to have. Let $K(q)$ be the set of $g(q)$ residue classes that are in $K(p)$ (permitted classes) for each prime factor p of q. At a/q we expect a spike of height proportional to

$$K(a,q) = \sum_{k \in K(q)} e_q(ak)/g(q). \tag{6.16}$$

If p is a repeated prime factor of q, then the classes $k+q/p$ are also in $K(q)$, and since replacement of k in eqn (6.16) by $k+q/p$ multiplies the right-hand side by $e_p(a)$, which is not unity, $K(a,q)$ is zero if q has any repeated prime factor. Now the simplest exponential sum is

$$F(\alpha) = \sum_{m=1}^{N} e(m\alpha), \tag{6.17}$$

which has a spike at $\alpha = 0$, since

$$|F(\alpha)| = |\sin \pi N\alpha / \sin \pi\alpha|. \tag{6.18}$$

We therefore compare $S(\alpha)$ with

$$T(\alpha) = \sum_{q \leqslant Q} \sideset{}{^*}\sum_{a \bmod q} K(a,q) F(\alpha - a/q). \tag{6.19}$$

The coefficient of $e(m\alpha)$ in $T(\alpha)$ is

$$\sum_{q \leqslant Q} \frac{1}{g(q)} \sum_{k \in K(q)} \sideset{}{^*}\sum_{a \bmod q} e_q(ak - am) = \sum_{q \leqslant Q} \frac{1}{g(q)} \sum_{k \in K(q)} c_q(k-m), \tag{6.20}$$

using the definition (1.11) of Ramanujan's sum. We know the sum over k must be zero if q has a repeated prime factor; if q is square-free then

$$\begin{aligned} \sum_{k \in K(q)} c_q(k-m) &= \sum_{k \in K(q)} \sum_{\substack{d|q \\ d|(k-m)}} d\mu\left(\frac{q}{d}\right) \\ &= \sum_{\substack{d|q \\ m \in K(d)}} d\mu\left(\frac{q}{d}\right) \frac{g(q)}{g(d)} \\ &= \mu(q) g(q) \mu(q_1) f(q_1)/g(q_1), \end{aligned} \tag{6.21}$$

where q_1 is the largest factor of q for which $m \in K(q_1)$. We write $q = q_1 q_2$, so that if $m \in H(d)$ then d divides q_2. The expression in (6.21) becomes

$$\mu(q_2) g(q_2) f(q)/f(q_2). \tag{6.22}$$

Now
$$\sum_{d|q_2} \frac{\mu(d)d}{f(d)} = \frac{\mu(q_2)g(q_2)}{f(q_2)}, \tag{6.23}$$
and so the coefficient of $e(m\alpha)$ in eqn (6.19) is
$$\sum_{\substack{d \leqslant Q \\ m \in H(d)}} \frac{\mu(d)d}{f(d)} \sum_{\substack{q \equiv 0 (\mathrm{mod}\, d) \\ q \leqslant Q}} \frac{\mu^2(q)f(q)}{g(q)}. \tag{6.24}$$

Selberg's argument proceeds as follows. If m is in $\mathcal{N}$, then $m \in H(d)$ only when $d = 1$, so that by eqn (6.4)
$$\int_0^1 S(\alpha)T(-\alpha) = \sum_{n_i \leqslant N} \sum_{q \leqslant Q} \frac{\mu^2(q)f(q)}{g(q)}. \tag{6.25}$$
We obtain (6.12) by Cauchy's inequality,
$$\left|\int_0^1 S(\alpha)T(-\alpha)\, \mathrm{d}\alpha\right|^2 \leqslant \int_0^1 |S(\alpha)|^2\, \mathrm{d}\alpha \int_0^1 |T(\beta)|^2\, \mathrm{d}\beta, \tag{6.26}$$
noting that, from eqn (6.4),
$$\int_0^1 |S(\alpha)|^2\, \mathrm{d}\alpha = \int_0^1 S(\alpha)S(-\alpha)\, \mathrm{d}\alpha = \sum_{n_i \leqslant N} 1, \tag{6.27}$$
and
$$\int_0^1 |T(\alpha)|^2\, \mathrm{d}\alpha = N \sum_{q \leqslant Q} \frac{\mu^2(q)f(q)}{g(q)} + \text{error term}. \tag{6.28}$$
We can prove eqn (6.28) either directly from (6.19) or by writing $|T(\alpha)|^2$ as $T(\alpha)T(-\alpha)$ and using (6.4) again and the expression (6.24) for the coefficients. Selberg used the second method, which can be expressed entirely in terms of the manipulation of the coefficients (6.24), with the integrations (6.25) and (6.28) occurring only implicitly. The first proof uses eqn (6.18) to show that the spikes of $T(\alpha)$ provide the main contribution to the integral in (6.28), and the observation that
$$\sum_{a \bmod q}^{*} |K(a,q)|^2 = \mu^2(q)f(q)/g(q). \tag{6.29}$$

The usual account of Selberg's method in terms of coefficient manipulations can be found in Halberstam and Roth (1966, Chapter 4). The sketch above is presented in terms of exponential sums for comparison with the 'large sieve' of the next two chapters.

7

THE LARGE SIEVE

But whatever his weight in pounds, shillings and ounces,
He always seems bigger because of his bounces.

II. 30

WE have seen that the behaviour of a given sequence of integers considered mod q is reflected in the behaviour of the sums $S(a/q)$ where

$$S(\alpha) = \sum_{1}^{N} a_m e(m\alpha), \tag{7.1}$$

in which a_m is 1 if m is in the given sequence, and 0 if m is not. An upper bound for the sum

$$\sum_{q \leqslant Q} \sum_{a \bmod q}^{*} |S(a/q)|^2 \tag{7.2}$$

(or for some related expression) is called a *large sieve* for residue classes. Other large sieves will appear in Chapter 18. In this chapter we prove what is probably the simplest of the upper bounds for the sum (7.2), and in the next chapter we shall use it to prove an upper bound of the form (6.12) for the number of elements of a sifted sequence in a bounded interval.

The proof does not require that a_m in eqn (7.1) takes only the values 0 or 1, and it treats (7.2) as a special case of the sum

$$\sum_{r=1}^{R} |S(x_r)|^2 \tag{7.3}$$

where $0 \leqslant x_1 < x_2 < \dots < x_R \leqslant 1$. Since a/q occurs in the sum (7.2) only if it is in its lowest terms, the points x_r are distinct in our proposed application. We write

$$\delta = \min\{x_2 - x_1, x_3 - x_2, \dots, x_1 + 1 - x_R\} \tag{7.4}$$

and suppose that $\delta > 0$. Before proving an upper bound for (7.3), we consider what form it might possibly take. Certainly there will exist sequences of coefficients a_m and points x_r for which

$$\sum_{r=1}^{R} |S(x_r)|^2 \geqslant \delta^{-1} \int_0^1 |S(\alpha)|^2 \, d\alpha$$

$$= \delta^{-1} \sum_{1}^{N} |a_m|^2. \tag{7.5}$$

On the other hand, any one term in the sum (7.3) may be as large as $(\sum |a_m|)^2$, and if all the a_m are equal in modulus this is

$$N \sum_1^N |a_m|^2. \tag{7.6}$$

An optimistic conjecture is that the inequality

$$\sum_{r=1}^{R} |S(x_r)|^2 \leqslant (N+\delta^{-1}) \sum_1^N |a_m|^2 \tag{7.7}$$

always holds. Surprisingly, the right-hand side of (7.7) has the correct order of magnitude: we shall prove that

$$\sum_{r=1}^{R} |S(x_r)|^2 \leqslant (N+\tfrac{2}{3}\delta^{-1}\sqrt{3}+O(1)) \sum_1^N |a_m|^2; \tag{7.8}$$

this result is due essentially to Bombieri (1972).

To prove the relation (7.8) we use the language of N-dimensional vectors over the complex numbers. The inner product $(\mathbf{g}, \mathbf{h})$ of two vectors $\mathbf{g} = (g_1,\dots, g_N)$ and $\mathbf{h} = (h_1,\dots, h_N)$ is given by

$$(\mathbf{g}, \mathbf{h}) = \sum_1^N g_n \bar{h}_n, \tag{7.9}$$

and the norm $\|\mathbf{g}\|$ by

$$\|\mathbf{g}\| = ((\mathbf{g}, \mathbf{g}))^{\frac{1}{2}}. \tag{7.10}$$

We can now state a fundamental lemma.

LEMMA. *Let* $\mathbf{u}, \mathbf{f}^{(1)},\dots, \mathbf{f}^{(R)}$ *be* N*-dimensional vectors, and* $c_1,\dots, c_R$ *be any complex coefficients. Then*

$$\left|\sum_{r=1}^{R} c_r(\mathbf{u}, \mathbf{f}^{(r)})\right| \leqslant \|\mathbf{u}\| \Big(\sum_1^R |c_r|^2\Big)^{\frac{1}{2}} \Big(\max_{r=1,\dots,R} \sum_{s=1}^{R} |(\mathbf{f}^{(r)}, \mathbf{f}^{(s)})|\Big)^{\frac{1}{2}}. \tag{7.11}$$

Proof. The left-hand side of (7.11) is

$$\left|\Big(\mathbf{u}, \sum_1^R c_r \mathbf{f}^{(r)}\Big)\right| \leqslant \|\mathbf{u}\| \times \left\|\sum_1^R \bar{c}_r \mathbf{f}^{(r)}\right\|. \tag{7.12}$$

The square of the second factor on the right-hand side of (7.12) is

$$\sum_{r=1}^{R} \sum_{s=1}^{R} \bar{c}_r c_s (\mathbf{f}^{(r)}, \mathbf{f}^{(s)}) \leqslant \tfrac{1}{2} \sum_{r=1}^{R} \sum_{s=1}^{R} (|c_r|^2 + |c_s|^2) |(\mathbf{f}^{(r)}, \mathbf{f}^{(s)})|$$

$$= \sum_{r=1}^{R} |c_r|^2 \sum_{s=1}^{R} |(\mathbf{f}^{(r)}, \mathbf{f}^{(s)})|, \tag{7.13}$$

from which the result follows.

In applying the lemma we choose c_r so that each summand on the left of (7.11) is real and positive. In Chapter 27 we shall apply the lemma with c_r of unit modulus to obtain Halász's method for estimating the number of times $S(\alpha)$ is large. To prove (7.8) we take c_r given by

$$\bar{c}_r = (\mathbf{f}^{(r)}, \mathbf{f}^{(s)}) \tag{7.14}$$

and deduce the corollary.

COROLLARY. *We have*

$$\sum_{r=1}^{R} |(\mathbf{u}, \mathbf{f}^{(r)})|^2 \leqslant \|\mathbf{u}\|^2 \max_{1 \leqslant r \leqslant R} \sum_{s=1}^{R} |(\mathbf{f}^{(r)}, \mathbf{f}^{(s)})|. \tag{7.15}$$

We write

$$S(\alpha) = e(M\alpha + \alpha) \sum_{-M}^{M} b_m e(m\alpha), \tag{7.16}$$

where $2M = N$ or $N-1$, whichever is even. We choose

$$f_m^{(r)} = k_m e(-mx_r), \tag{7.17}$$

$$u_m = b_m / k_m, \tag{7.18}$$

where $k_{-M}, \ldots, k_M$ are given by

$$\left.\begin{aligned} k_m &= 1 \quad \text{for } |m| \leqslant M \\ k_m &= \left(1 - \frac{|m| - M}{L}\right)^{\frac{1}{2}} \quad \text{for } M \leqslant |m| \leqslant M + L \\ k_m &= 0 \quad \text{for } |m| \geqslant M + L \end{aligned}\right\}; \tag{7.19}$$

here L is a parameter which we shall choose below. We now have

$$|(\mathbf{u}, \mathbf{f}^{(r)})| = \left|\sum_{-M}^{M} b_m e(mx_r)\right| = |S(x_r)|. \tag{7.20}$$

Moreover

$$\|\mathbf{u}\|^2 = \sum_{-M}^{M} |b_m|^2 / k_m^2 = \sum_{1}^{N} |a_m|^2 \tag{7.21}$$

and

$$(\mathbf{f}^{(r)}, \mathbf{f}^{(s)}) = K(x_s - x_r), \tag{7.22}$$

where

$$\begin{aligned} K(\alpha) &= \sum_{-M-L}^{M+L} k_m^2 e(m\alpha) \\ &= \frac{\sin^2(M+L)\pi\alpha - \sin^2 M\pi\alpha}{L \sin^2 \pi\alpha}. \end{aligned} \tag{7.23}$$

Hence we have

$$\begin{aligned} \sum_{s=1}^{R} |K(x_s - x_r)| &\leqslant 2M + L + 2 \sum_{1}^{\frac{1}{2}(R-1)} L^{-1} \operatorname{cosec}^2 \pi m\delta \\ &\leqslant 2M + L + 2L^{-1}\pi^2 \sum_{1}^{\infty} m^{-2}\delta^{-2} + O(L^{-1}) \\ &\leqslant 2M + L + (3L\delta^2)^{-1} + O(L^{-1}). \end{aligned} \tag{7.24}$$

When we choose L to be an integer close to $(\delta\sqrt{3})^{-1}$, (7.24) becomes

$$\leqslant 2M+\tfrac{2}{3}\delta^{-1}\sqrt{3}+O(1). \tag{7.25}$$

We substitute (7.20), (7.21), and (7.25) into (7.15) to obtain (7.8).

Our inequality (7.8) represents an improvement of an inequality of Roth (1965), which has led to much recent work. The best upper bounds known for the sum (7.3) at the time of writing are

$$N\left(1+\frac{2\sqrt{3}}{3N\delta}+\frac{3}{N}\right)\sum_{1}^{N}|a_m|^2, \tag{7.26}$$

$$\delta^{-1}(1+270N^3\delta^3)\sum_{1}^{N}|a_m|^2, \tag{7.27}$$

and

$$2\max(N,\delta^{-1})\sum_{1}^{N}|a_m|^2. \tag{7.28}$$

Of these, (7.26) is the result of this chapter, appropriate when $N > \frac{2}{3}\delta^{-1}\sqrt{3}$, and (7.27) and (7.28) are results of Bombieri and Davenport (1969, 1968). (7.27) is appropriate when $\delta^{-1} > 3(10)^{\frac{1}{3}}N$, and (7.28) for the intermediate range.

Note added in proof. H. Montgomery and R. C. Vaughan have now proved the conjecture (7.7). This supersedes (7.26) and (7.28) but not (7.27).

8

THE UPPER-BOUND SIEVE

'It's a comfortable sort of thing to have', said Christopher Robin, folding up the paper and putting it into his pocket.
II. 170

In this chapter we obtain the upper bound (6.12) as an application of the large sieve. The notation is that of Chapter 6. $\mathscr{N}$ is a sequence of positive integers, and for each $q \leqslant Q$ there are sets $H(q)$ and $K(q)$ of residue classes $\bmod q$. The $f(q)$ classes of $H(q)$ are precisely those that are not congruent to any member of $\mathscr{N} \bmod p$ for any prime p dividing q, so that, if h is in a class of $H(q)$ and $n \in \mathscr{N}$,

$$(n-h, q) = 1. \tag{8.1}$$

The $g(q)$ classes of $K(q)$ are those that for each p dividing q are congruent $\bmod p$ to some member of $\mathscr{N}$; their union contains all members of the sequence $\mathscr{N}$.

We work with the exponential sum of eqn (6.15):

$$S(\alpha) = \sum_{n_i \leqslant N} e(n_i \alpha), \tag{8.2}$$

where the sum is over members $n_1, n_2, \ldots$ of the sequence $\mathscr{N}$. If h is a class of $H(q)$,

$$\begin{aligned} \sum_{a \bmod q}^{*} S(a/q) e_q(-ah) &= \sum_{n_i \leqslant N} c_q(n_i - h) \\ &= \mu(q) M, \end{aligned} \tag{8.3}$$

where M is the number of members of $\mathscr{N}$, and we have used eqn (3.19) in the special case (8.1). Hence

$$\mu(q) f(q) M = \sum_{a \bmod q}^{*} \sum_{h \in H(q)} S(a/q) e_q(-ah). \tag{8.4}$$

Cauchy's inequality now gives

$$\mu^2(q) f^2(q) M^2 \leqslant \Big(\sum_{a \bmod q}^{*} |S(a/q)|^2 \Big) \Big(\sum_{a \bmod q} \Big| \sum_{h \in H(q)} e_q(-ah) \Big|^2 \Big). \tag{8.5}$$

The second sum over a on the right-hand side of (8.5) can be rearranged as follows.

$$\sideset{}{^*}\sum_{a \bmod q} \sum_{h \in H(q)} e_q(-ah) \sum_{g \in H(q)} e_q(ag) = \sum_{g \in H(q)} \sum_{h \in H(q)} c_q(g-h)$$

$$= \sum_{g \in H(q)} \sum_{h \in H(q)} \sum_{\substack{d \mid q \\ d \mid (g-h)}} d\mu(q/d)$$

$$= \sum_{d \mid q} d\mu(q/d) f^2(q)/f(d). \tag{8.6}$$

In eqn (8.6) we have a sum that is a Möbius inverse (in the sense of (1.27)) of that in

$$\sum_{r \mid m} \frac{\mu^2(r)g(r)}{f(r)} = \frac{m}{f(m)}, \tag{8.7}$$

an equation that expresses the fact that for a prime modulus p, every residue class is either in $H(p)$ or in $K(p)$. The terms involving d in (8.6) therefore come to $\mu^2(q)g(q)/f(q)$, and we can put (8.5) into the form

$$\frac{\mu^2(q)f(q)}{g(q)} M^2 \leqslant \sideset{}{^*}\sum_{a \bmod q} \left| S\left(\frac{a}{q}\right) \right|^2. \tag{8.8}$$

We apply the large sieve (7.8) with the rationals a/q with $q \leqslant Q$ and $(a, q) = 1$ as the points $x_1, \ldots, x_R$, so that

$$\delta = (Q(Q-1))^{-1} \tag{8.9}$$

in eqn (7.4). The upper bound (7.8) now gives

$$\sum_{q \leqslant Q} \sideset{}{^*}\sum_{a \bmod q} |S(a/q)|^2 \leqslant (N + O(Q^2))M. \tag{8.10}$$

Combining (8.8) and (8.10), we have

$$M^2 \sum_{q \leqslant Q} \mu^2(q)f(q)/g(q) \leqslant (N + O(Q^2))M, \tag{8.11}$$

or

$$M \leqslant \frac{N + O(Q^2)}{\sum_{q \leqslant Q} \mu^2(q)f(q)/g(q)}, \tag{8.12}$$

which is the relation (6.12) with an explicit error term. We have stated our result for the interval $1 \leqslant n \leqslant N$ of the sequence $\mathcal{N}$, but (8.12) holds as an upper bound for the number of members of $\mathcal{N}$ in any interval $L+1 \leqslant n \leqslant L+N$; we have to consider merely the integers $n-L$, where n is in $\mathcal{N}$, themselves a sifted sequence with the same values of $f(q)$ and $g(q)$ but the sets $H(q)$ and $K(q)$ translated by L. The deduction of (8.12) from (7.8) is due to Montgomery (1968). His argument differs from that given here; there are several alternative proofs.

Two worked examples follow. First we consider the perfect squares not exceeding N. These are a sifted sequence: the set $H(p)$ contains the residue classes mod p that do not contain squares, and thus

$$\left.\begin{array}{ll} f(2) = 0, & g(2) = 2 \\ f(p) = \tfrac{1}{2}(p-1), & g(p) = \tfrac{1}{2}(p+1) \quad \text{for } p \geqslant 3 \end{array}\right\}. \tag{8.13}$$

It is not difficult to show that

$$\sum_{q \leqslant Q} \mu^2(q) f(q)/g(q) = (c+o(1))Q, \tag{8.14}$$

where c is a constant, as $Q \to \infty$. Choosing $Q = N^{\frac{1}{2}}$, we have shown that the number of perfect squares not exceeding N is $O(N^{\frac{1}{2}})$. This is very encouraging, since the sieve upper bound is 'sharp', differing only by a constant factor from the actual number of squares. It is surprising that we have not lost the correct order of magnitude in combining so many inequalities.

Our second, less trivial example concerns the primes between Q and N; these form a sifted sequence with

$$f(p) = 1, \qquad g(p) = p-1 = \varphi(p), \tag{8.15}$$

and thus

$$\begin{aligned} \sum_{q \leqslant Q} \frac{\mu^2(q) f(q)}{g(q)} &= \sum_{q \leqslant Q} \frac{\mu^2(q)}{\varphi(q)} \\ &= (1+o(1)) \log Q, \end{aligned} \tag{8.16}$$

when worked out as in Chapter 2. We can obtain a lower bound for the sum in (8.16) with less effort, since

$$\begin{aligned} \sum_{q \leqslant Q} \frac{\mu^2(q)}{\varphi(q)} &= \sum_{q \leqslant Q} \frac{\mu^2(q)}{q} \prod_{p|q} \left(1+\frac{1}{p}+\frac{1}{p^2}+\dots\right) \\ &= \sum 1/m, \end{aligned} \tag{8.17}$$

the sum being over all m whose prime factors are at most Q. All $m \leqslant Q$ are included in this sum, and thus

$$\sum_{q \leqslant Q} \frac{\mu^2(q)}{\varphi(q)} > \sum_{m \leqslant Q} \frac{1}{m} > \log Q \tag{8.18}$$

for $q > 1$, by (2.5). When we choose Q a little smaller than $N^{\frac{1}{2}}$ we see that the primes between 1 and N number

$$\leqslant Q + (N + O(Q^2))/\log Q \leqslant (2+o(1))N/\log N. \tag{8.19}$$

The right-hand side of (8.19) is just double the true value (5.23). Moreover, (8.19) is also an upper bound for the number of primes in any interval of length N.

We derived the inequality (8.12) from Cauchy's inequality; the difference between the two sides of the inequality (8.12) is a measure of how closely the values of $S(a/q)$ are proportional to those of

$$(g(q))^{-1} \sum_{k \in K(q)} e_q(ak), \tag{8.20}$$

and this in its turn measures how evenly $\mathcal{N}$ is distributed among the $g(q)$ residue classes mod q into which it is allowed to fall. We could add an explicit term on the right of (8.8) to measure the unevenness (what statisticians might term a variance). The inequality (7.8) gives a strong upper bound for this variance as well as for the main term. When we use (7.8) to prove Bombieri's theorem, it is the variance bound that is important, not the bound for the main term.

9

FRANEL'S THEOREM

'It's just a thing you discover', said Christopher Robin carelessly, not being quite sure himself.

I. 109

THE *Farey sequence of order Q* consists of the fractions a/q in their lowest terms (i.e. $(a,q) = 1$), with $q \leqslant Q$ and $0 < a \leqslant q$. We name them $f_r = a_r/q_r$ in increasing order, so that $f_1 = 1/Q, f_2 = 1/(Q-1),..., f_F = 1$. For notational convenience we may refer to f_{F+r}; this is to be interpreted as $1+f_r$. Here F is the number of terms in the Farey sequence, so that

$$F = \sum_{q \leqslant Q} \varphi(q) = 3\pi^{-2}Q^2 + O(Q \log Q) \tag{9.1}$$

from eqn (2.11). The properties of the Farey sequence are discussed by Hardy and Wright (1960, Chapter 3).

We shall sketch a proof that

$$f_{r+1} - f_r = (q_r q_{r+1})^{-1}. \tag{9.2}$$

Let us represent rational numbers a/q (not necessarily in their lowest terms) by points (a, q) of two-dimensional Euclidean space. Since f_r and f_{r+1} are consecutive, the only integer points in the closed triangle with vertices O $(0, 0)$, P_r (a_r, q_r), and P_{r+1} (a_{r+1}, q_{r+1}) are its three vertices. By symmetry, the only integer points in the parallelogram $OP_r TP_{r+1}$ are its vertices, where T is $(a_r + a_{r+1}, q_r + q_{r+1})$. We can now cover the plane with the translations of this parallelogram in such a way that integer points occur only at the vertices of parallelograms. It follows that $OP_r TP_{r+1}$ has unit area, which is the assertion (9.2).

Before stating Franel's theorem we introduce some notation. For $0 < \alpha \leqslant 1$ we write

$$E(\alpha) = \sum_{f_r \leqslant \alpha} 1 - \alpha F, \tag{9.3}$$

so that $E(\alpha)$ is the excess number of Farey fractions in $(0, \alpha]$ beyond the expected number αF. Franel considered the sum

$$\sum_{r=1}^{F} |E(f_r)|^2 \tag{9.4}$$

and showed that it is $o(Q^4)$ if and only if eqn (5.18) holds; and an upper bound Q^λ for (9.4) with $3 < \lambda < 4$ is valid if and only if

$$|M(x)| \ll x^{\frac{1}{2}\lambda - 1}. \tag{9.5}$$

We can also connect $M(x)$ with

$$\int_0^1 |E(\alpha)|^2 \, d\alpha. \tag{9.6}$$

In fact, the ratio of (9.4) and (9.6) lies between bounded multiples of F; but this fact requires proof, as the Riemann sum corresponding to the integral (9.6) and the points $f_1, \ldots, f_F$ is

$$\sum_{r=1}^{F} (f_{r+1} - f_r) |E(f_r)|^2, \tag{9.7}$$

and from eqn (9.2) we see that the difference $f_{r+1} - f_r$ varies from Q^{-1} almost down to Q^{-2}. Franel (1924) produced a curious identity for the sum (9.4). We shall show in (9.20) that (9.4) is less than a bounded multiple of Franel's expression involving $M(x)$, and deduce the 'only if' clause of Franel's result by a method of Landau (1927, Vol. II, pp. 169–77).

We use the function $H(\alpha)$ of eqns (4.5) and (4.6):

$$H(\alpha) = \begin{cases} \alpha - [\alpha] - \frac{1}{2} & \text{if } \alpha \text{ is not an integer,} \\ 0 & \text{if } \alpha \text{ is an integer.} \end{cases} \tag{9.8}$$

We have

$$-\sum_{r=1}^{F} H(\alpha - f_r) = \begin{cases} E(\alpha) & \text{if } \alpha \text{ is in the Farey sequence,} \\ E(\alpha) - \frac{1}{2} & \text{if not,} \end{cases} \tag{9.9}$$

for $0 < \alpha \leqslant 1$. To verify eqn (9.9), we observe that the left-hand side is αF plus a step function that is zero at $\alpha = 0$ and has discontinuities $-\frac{1}{2}$ on each side of the Farey points f_r. Taking (4.7) in the form

$$\left| H(\alpha) + \sum_{\substack{m=-Q^2 \\ m \neq 0}}^{Q^2} \frac{e(m\alpha)}{2\pi i m} \right| \ll \frac{1}{Q^2 \|\alpha\|}, \tag{9.10}$$

we have from (9.9)

$$\begin{aligned} E(f_r) &= -\sum_{t=1}^{F} H(f_r - f_t) \\ &= \sum_{\substack{t=1 \\ t \neq r}}^{F} \left\{ \sum_{\substack{m=-Q^2 \\ m \neq 0}}^{Q^2} \frac{e(mf_r - mf_t)}{2\pi i m} + O\left(\frac{1}{Q^2 \|f_r - f_t\|} \right) \right\}. \end{aligned} \tag{9.11}$$

By (9.2), the consecutive Farey fractions are at least Q^{-2} apart, and so the sum of the error terms in (9.11) is

$$\ll Q^{-2}\sum_{t=1}^{F} Q^2/t \ll \log F \ll \log Q, \tag{9.12}$$

where we have used expressions (2.5) and (9.1). We can now replace the term $t = r$ (since $H(0) = 0$) and rearrange the first term on the right of (9.11) as

$$\sum_{t=1}^{F} \sum_{\substack{m=-Q^2 \\ m\neq 0}}^{Q^2} (2\pi i m)^{-1} e(mf_r - mf_t) = \sum_{\substack{-Q^2 \\ m\neq 0}}^{Q^2} (2\pi i m)^{-1} \sum_{q\leqslant Q} \sum_{a \bmod q}^{*} e(mf_r)e_q(-am)$$

$$= \sum_{\substack{-Q^2 \\ m\neq 0}}^{Q^2} (2\pi i m)^{-1} e(mf_r) \sum_{q\leqslant 0} c_q(-m). \tag{9.13}$$

We apply the large sieve (7.8) with (9.13) as the exponential sum and $f_1,\ldots, f_F$ as the points. By eqn (9.2), δ^{-1} is here $Q(Q-1)$. We have

$$\sum_{r=1}^{F} \left| \sum_{\substack{-Q^2 \\ m\neq 0}}^{Q^2} (2\pi i m)^{-1} e(mf_r) \sum_{q\leqslant Q} c_q(-m)\right|^2 \ll Q^2 \sum_{\substack{-Q^2 \\ m\neq 0}}^{Q^2} m^{-2} \left|\sum_{q\leqslant 0} c_q(-m)\right|^2$$

$$\ll Q^2 \sum_{m=1}^{Q^2} m^{-2} \left| \sum_{d|m} d \sum_{\substack{q\leqslant Q \\ q\equiv 0 (\bmod d)}} \mu(q/d)\right|^2, \tag{9.14}$$

where we have used eqn (3.19) for Ramanujan's sum. The coefficient of Q^2 in (9.14) is now

$$\leqslant \sum_{d\leqslant Q} dM(Q/d) \sum_{f\leqslant Q} fM(Q/f) \sum_{\substack{m\equiv 0 (\bmod d) \\ m\equiv 0 (\bmod f)}} m^{-2}, \tag{9.15}$$

where we have taken the sum over m from $m = 1$ to infinity, since the coefficient of m^{-2} in (9.14) is non-negative. The inner sum in (9.15) is

$$\frac{\pi^2}{6}\frac{1}{[d,f]^2} = \frac{\pi^2}{6}\frac{(d,f)^2}{d^2f^2}. \tag{9.16}$$

Let $u(m)$ be the arithmetical function with

$$\sum_{d|m} u(d) = m^2. \tag{9.17}$$

By Möbius's inversion (1.28), we can verify that

$$0 \leqslant u(t) \leqslant t^2. \tag{9.18}$$

We can now write (9.15) as

$$\tfrac{1}{6}\pi^2 \sum_{d\leqslant Q} dM(Q/d) \sum_{f\leqslant Q} fM(Q/f) d^{-2} f^{-2} \sum_{\substack{t|d \\ t|f}} u(t) \ll \sum_{t\leqslant Q} u(t) \Big(\sum_{\substack{d\leqslant Q \\ d\equiv 0 (\bmod t)}} d^{-1} M(Q/d)\Big)^2. \tag{9.19}$$

Squaring (9.11) and substituting (9.12) and (9.19), we have

$$\sum_{r=1}^{F} |E(f_r)|^2 \ll Q^2 \sum_{t \leqslant Q} u(t)\Big(\sum_{\substack{d \leqslant Q \\ d \equiv 0 (\bmod t)}} d^{-1}M(Q/d)\Big)^2 + F\log^2 Q$$

$$\ll Q^2 \sum_{t \leqslant Q} t^2\Big(\sum_{\substack{d \leqslant Q \\ d \equiv 0 (\bmod t)}} d^{-1}M(Q/d)\Big)^2 + Q^2\log^2 Q. \qquad (9.20)$$

The leading term on the right of (9.20) is essentially the sum involving Möbius functions of Franel's identity.

The inequality from (9.8) to $M(x)$ is less involved. Since, by (3.19),

$$\sum_{a \bmod q}^{*} e_q(a) = c_q(1) = \mu(q), \qquad (9.21)$$

we have

$$M(Q) = \sum_{r=1}^{F} e(f_r). \qquad (9.22)$$

Now

$$|e(f_r) - e(r/F)| \leqslant 2\pi |f_r - r/F|$$
$$\leqslant 2\pi F^{-1}|E(f_r)|. \qquad (9.23)$$

Since

$$\sum_{r=1}^{F} e(r/F) = 0 \qquad (9.24)$$

from (3.1), we have

$$|M(Q)| \leqslant 2\pi F^{-1} \sum_{r=1}^{F} |E(f_r)|$$
$$\ll F^{-1}F^{\frac{1}{2}}\Big(\sum_{r=1}^{F} |E(f_r)|^2\Big)^{\frac{1}{2}}$$
$$\ll Q^{-1}\Big(\sum_{r=1}^{F} |E(f_r)|^2\Big)^{\frac{1}{2}}, \qquad (9.25)$$

and so a bound for (9.8) of the form $o(Q^4)$ implies the prime-number theorem in the form (5.18).

PART II

The Prime-Number Theorem

10

A MODULAR RELATION

> Winnie-the-Pooh read the two notices very carefully, first from left to right, and afterwards, in case he had missed some of it, from right to left.
>
> I. 46

IN this second part we study $\zeta(s)$ and $L(s, \chi)$ as functions of the complex variable s, and work towards the prime-number theorem. Our investigations are based on the functional equations for $\zeta(s)$ and for $L(s, \chi)$. The first step is therefore to prove these.

We need a lemma from the theory of Fourier series.

LEMMA. (*Poisson's summation formula.*) *Let k, l be integers. Let $f(x)$ be a differentiable function of a real variable with*

$$|f'(x)| \leqslant A$$

on $[k,l]$. Then

$$\sum_{m=-\infty}^{\infty} \int_{k}^{k+l} f(x)e(mx)\,dx = \tfrac{1}{2}f(k)+f(k+1)+\ldots+f(k+l-1)+\tfrac{1}{2}f(k+l) \tag{10.1}$$

and moreover the partial sums satisfy the inequality

$$\left|\sum_{-M}^{M} \int_{k}^{k+l} f(x)e(mx)\,dx-\tfrac{1}{2}f(k)-\ldots-\tfrac{1}{2}f(k+l)\right| \leqslant lAM^{-1}\log M. \tag{10.2}$$

Proof. Equation (10.1) is additive on intervals: its truth for $[k,r]$ and for $[r,t]$ implies its truth for $[k,t]$; so we may suppose $l = 1$. By a change of variable we may suppose also $k = 0$. We use the function $H(x)$ of eqn (4.4):

$$H(x) = \sum_{\substack{m=-\infty \\ m\neq 0}}^{\infty} \frac{e(mx)}{-2\pi i m} = \begin{cases} x-\frac{1}{2} & \text{if } 0 < x < 1, \\ 0 & \text{if } x = 0 \text{ or } 1, \end{cases} \tag{10.3}$$

and in particular if $0 < x < 1$ we have (4.6):

$$\left| H(x) + \sum_{\substack{-M \\ m\neq 0}}^{M} \frac{e(mx)}{2\pi i m} \right| \ll (M\|x\|)^{-1}. \tag{10.4}$$

Hence

$$\int_0^1 \left| H(x) + \sum_{\substack{-M \\ m\neq 0}}^{M} (2\pi i m)^{-1} e(mx) \right| \mathrm{d}x \ll \int_0^{1/M} \sum_1^M m^{-1}\,\mathrm{d}x + \int_{1/M}^{1/2} M^{-1}x^{-1}\,\mathrm{d}x$$
$$\ll M^{-1}\log M. \tag{10.5}$$

We now have

$$\sum_{\substack{-M \\ m\neq 0}}^{M} \int_0^1 f(x)e(mx)\,\mathrm{d}x$$
$$= \left[f(x) \sum_{\substack{-M \\ m\neq 0}}^{M} (2\pi i m)^{-1} e(mx) \right]_0^1 - \int_0^1 f'(x) \sum_{\substack{-M \\ m\neq 0}}^{M} (2\pi i m)^{-1} e(mx)\,\mathrm{d}x. \tag{10.6}$$

The first term on the right of eqn (10.6) is zero, since the terms for m and $-m$ cancel, and by (10.5) the second is

$$\int_0^1 f'(x)H(x)\,\mathrm{d}x + O(AM^{-1}\log M). \tag{10.7}$$

The integral in (10.7) is

$$\int_0^1 f'(x)(x-\tfrac{1}{2})\,\mathrm{d}x = \tfrac{1}{2}f(0) + \tfrac{1}{2}f(1) - \int_0^1 f(x)\,\mathrm{d}x. \tag{10.8}$$

Combining (10.6), (10.7), and (10.8), we have

$$\sum_{-M}^{M} \int_0^1 f(x)e(mx)\,\mathrm{d}x = \tfrac{1}{2}f(0) + \tfrac{1}{2}f(1) + O(AM^{-1}\log M), \tag{10.9}$$

and letting M tend to infinity we have the case $k = 0, l = 1$ of (10.1).

A general method of proving functional equations is to write the required function as an infinite series, apply Poisson's summation formula to a partial sum, and then let the length of the sum tend to infinity. This entails a change in the order of summation in a double infinite series. We could prove the functional equations for $\zeta(s)$ and

$L(s, \chi)$ directly by this method, but it is more troublesome to justify the interchange of summations and more difficult to identify the functions that arise. We shall therefore prove the identity

$$\sum_{m=-\infty}^{\infty} \exp\{-(m+\delta)^2\pi x^{-1}\} = x^{\frac{1}{2}} \sum_{m=-\infty}^{\infty} \exp(-m^2\pi x)e(m\delta), \tag{10.10}$$

where x is real and $0 \leqslant \delta \leqslant 1$. Rather than apply (10.2) directly, we use eqn (10.9), so that

$$\sum_{-M}^{M} \int_0^1 \exp\{-(r+t+\delta)^2\pi x^{-1}\}\, \mathrm{d}t$$
$$= \tfrac{1}{2}\exp\{-(r+\delta)^2\}+\tfrac{1}{2}\exp\{-(r+1+\delta)^2\}+$$
$$+O[rx^{-1}\exp\{-(r+\delta)^2\pi x^{-1}\}M^{-1}\log M], \tag{10.11}$$

and the double series

$$\sum_{m=-\infty}^{\infty} \sum_{r=-\infty}^{\infty} \int_r^{r+1} \exp\{-(t+\delta)^2\pi x^{-1}\}e(mt)\, \mathrm{d}t \tag{10.12}$$

converges and is equal to

$$\sum_{r=-\infty}^{\infty} \exp\{-(r+\delta)^2\pi x^{-1}\}. \tag{10.13}$$

A typical term in (10.12) is

$$\int_{-\infty}^{\infty} \exp\{-(t+\delta)^2\pi x^{-1}+2\pi \mathrm{i} mt\}\, \mathrm{d}t$$
$$= \int_{-\infty}^{\infty} \exp\{-\pi xz^2+2\pi \mathrm{i} m(xz-\delta)\}x\, \mathrm{d}z,$$
$$= x\exp\{-\pi xm^2-2\pi \mathrm{i} m\delta\} \int_{-\infty}^{\infty} \exp\{-\pi x(z-\mathrm{i} m)^2\}\, \mathrm{d}z, \tag{10.14}$$

where we have put $t+\delta = xz$. The integral in (10.14) can naturally be regarded as the contour integral of e^{-z^2} along the line $\operatorname{Im} z = m$. To evaluate the integral, we can move the line of integration to the real axis, when (10.14) becomes

$$x\exp(-\pi xm^2-2\pi \mathrm{i} m\delta) \int_{-\infty}^{\infty} \exp(-\pi xt^2)\, \mathrm{d}t$$
$$= c\pi^{-\frac{1}{2}}x^{\frac{1}{2}}\exp(-\pi xm^2-2\pi \mathrm{i} m\delta), \tag{10.15}$$

where
$$c = \int_{-\infty}^{\infty} \exp(-t^2)\, \mathrm{d}t. \tag{10.16}$$

If we combine (10.15) and (10.12) we have proved that

$$\sum_{r=-\infty}^{\infty} \exp\{-(r+\delta)^2\pi x^{-1}\} = \sum_{m=-\infty}^{\infty} c\pi^{-\frac{1}{2}}x^{\frac{1}{2}}\exp(-\pi m^2 x)e(m\delta). \quad (10.17)$$

Putting $\delta = 0$, $x = 1$ verifies that

$$c = \pi^{\frac{1}{2}}, \quad (10.18)$$

and we have proved (10.10).

Equation (10.17) should not be let pass without some comment. Let us put

$$\theta(\omega) = \sum_{m=-\infty}^{\infty} \exp(\pi \mathrm{i} m^2 \omega) \quad (10.19)$$

for values of ω for which the series in (10.19) converges, that is, for complex ω with positive real part. Then we have

$$\theta(\omega+2) = \theta(\omega), \quad (10.20)$$

and (10.10) with $\delta = 0$ gives us

$$\theta^2(-1/\omega) = \omega\theta^2(\omega), \quad (10.21)$$

provided ω is pure imaginary. However, since eqn (10.21) holds along the imaginary axis, the two sides of (10.21) have the same derivatives at points $\mathrm{i}y$ with $y > 0$, and, since power-series expansions of regular functions are unique, (10.21) must hold whenever $\theta(\omega)$ and $\theta(-1/\omega)$ are both defined, which is whenever the imaginary part of ω is positive. From eqns (10.20) and (10.21) we see that

$$\theta^4(\omega)\, \mathrm{d}\omega \quad (10.22)$$

is invariant under the group of transformations of the upper half of the complex plane generated by $\omega \to \omega+2$ and $\omega \to -1/\omega$.

We are now in the realm of the elliptic modular functions. A *modular function* is one that is invariant under the group of transformations generated by $\omega \to \omega+1$ and $\omega \to -1/\omega$ or under a subgroup of finite index in this group. The derivatives of a modular function are not invariant under these transformations, since $\mathrm{d}\omega$ itself is not invariant; functions that satisfy the transformation law for a power of a derivative of a modular function are called *modular forms*. The name 'elliptic modular functions' arises as follows. The periods of an elliptic function form a free Abelian group on two generators ω_1 and ω_2. A modular form corresponds to a function of two complex variables ω_1 and ω_2 which is homogeneous and whose value does not change when we replace ω_1 and ω_2 by another pair of generators of the same free Abelian group (or, more generally, which takes a finite set of different values when ω_1

and ω_2 are generators of the same Abelian group). Here $\omega = \omega_2/\omega_1$. Thus we may consider

$$f(\omega_1, \omega_2) = \omega_1^{-1}\theta^2(\omega_2/\omega_1), \tag{10.23}$$

with

$$f(\omega_2, -\omega_1) = \omega_2^{-1}\theta^2(-\omega_1/\omega_2) = \omega_1^{-1}\theta^2(\omega_2/\omega_1) = f(\omega_1, \omega_2) \tag{10.24}$$

and

$$f(\omega_1, \omega_2+2\omega_1) = \omega_1^{-1}\theta^2(\omega_2/\omega_1+2) = f(\omega_1, \omega_2). \tag{10.25}$$

11

THE FUNCTIONAL EQUATIONS

When he awoke in the morning, the first thing he saw was Tigger, sitting in front of the glass and looking at himself.

'Hallo!' said Pooh.

'Hallo!' said Tigger. 'I've found somebody just like me. I thought I was the only one of them.' II. 21

WE return to $\zeta(s)$ and $L(s, \chi)$. The definition

$$\Gamma(\tfrac{1}{2}s) = \int_0^\infty e^{-y} y^{\frac{1}{2}s-1}\,dy, \tag{11.1}$$

valid for $\sigma > 0$ (we recall $s = \sigma + it$, σ and t real), becomes

$$\Gamma(\tfrac{1}{2}s) = \pi^{\frac{1}{2}s} m^s \int_0^\infty e^{-m^2\pi x} x^{\frac{1}{2}s-1}\,dx \tag{11.2}$$

when we write $y = \pi m^2 x$. Summation over m gives

$$\pi^{-\frac{1}{2}s}\Gamma(\tfrac{1}{2}s)\zeta(s) = \sum_{m=1}^{\infty} \int_0^\infty e^{-m^2\pi x} x^{\frac{1}{2}s-1}\,dx. \tag{11.3}$$

Since the sum and integral in eqn (11.3) each converge absolutely, we can rearrange the right-hand side of (11.3) as

$$\int_0^\infty \sum_1^\infty e^{-m^2\pi x} x^{\frac{1}{2}s-1}\,dx = \int_0^\infty \tfrac{1}{2}(\theta(ix)-1) x^{\frac{1}{2}s-1}\,dx, \tag{11.4}$$

where $\theta(\omega)$ is the function of eqn (10.19). Next we write

$$\int_0^1 \tfrac{1}{2}(\theta(ix)-1) x^{\frac{1}{2}s-1}\,dx = \int_1^\infty (\theta(i/t)-1) t^{-\frac{1}{2}s-1}\,dt \tag{11.5}$$

and use eqn (10.10) to put (11.5) into the form

$$\begin{aligned} \int_1^\infty \tfrac{1}{2}(t^{\frac{1}{2}}\theta(it)-1) t^{-\frac{1}{2}s-1}\,dt &= \int_1^\infty \tfrac{1}{2}\theta(it) t^{-\frac{1}{2}s-\frac{1}{2}}\,dt - \tfrac{1}{2}(2s^{-1}) \\ &= \int_1^\infty \tfrac{1}{2}(\theta(it)-1) t^{-\frac{1}{2}s-\frac{1}{2}}\,dt - s^{-1} - (1-s)^{-1}. \end{aligned} \tag{11.6}$$

The left-hand side of (11.3) has now been expressed as

$$\int_1^\infty \tfrac{1}{2}(\theta(\mathrm{i}x)-1)(x^{\frac{1}{2}s-1}+x^{-\frac{1}{2}s-\frac{1}{2}})\,\mathrm{d}x-\{s(1-s)\}^{-1}. \tag{11.7}$$

The expression (11.7) was obtained under the assumption $\sigma > 1$, but, since

$$\theta(\mathrm{i}x)-1 \ll \mathrm{e}^{-\pi x} \tag{11.8}$$

for $x > 1$, the integral in (11.7) converges for all complex s. Since $\Gamma(\frac{1}{2}s)$ is a known function, and $(\Gamma(\frac{1}{2}s))^{-1}$ is integral (single-valued and regular over the whole s-plane), we can take (11.3) with (11.7) as the definition of $\zeta(s)$, knowing that $\sum_1^\infty m^{-s}$ agrees with our new definition when the series converges. We have now continued $\zeta(s)$ over the whole plane. Further, (11.7) is unchanged when we replace s by $1-s$, so that

$$\pi^{-\frac{1}{2}s}\Gamma(\tfrac{1}{2}s)\zeta(s) = \pi^{\frac{1}{2}s-\frac{1}{2}}\Gamma(\tfrac{1}{2}-\tfrac{1}{2}s)\zeta(1-s), \tag{11.9}$$

the promised functional equation. Since

$$\frac{\Gamma(\frac{1}{2}s)}{\Gamma(\frac{1}{2}-\frac{1}{2}s)} = 2^{1-s}\pi^{-\frac{1}{2}}\Gamma(s)\cos\tfrac{1}{2}s\pi, \tag{11.10}$$

an alternative form of (11.9) is

$$\zeta(1-s) = 2^{1-s}\pi^{-s}\Gamma(s)\cos\tfrac{1}{2}s\pi\,\zeta(s). \tag{11.11}$$

We now list some properties of $\Gamma(s)$ (see for example Jeffreys and Jeffreys 1962, Chapter 15). The product

$$\frac{1}{\Gamma(s+1)} = \mathrm{e}^{\gamma s}\prod_{m=1}^{\infty}\left(1+\frac{s}{m}\right)\mathrm{e}^{-s/m}, \tag{11.12}$$

where γ is the constant of (2.5), converges for all s, and defines $\Gamma(s)$ as a function that is never zero and has simple poles at $0, -1, -2, \ldots$. Using this information in (11.7) we see that the pole of (11.7) at 1 comes from $\zeta(s)$, the pole at 0 from $\Gamma(\frac{1}{2}s)$, and that $\zeta(s)$ must have zeros at $s = -2, -4, \ldots$, to cancel the other poles of $\Gamma(\frac{1}{2}s)$. From eqn (11.12),

$$\Gamma(s+1) = s\Gamma(s), \tag{11.13}$$

and

$$\Gamma(1+s)\Gamma(1-s) = \pi s \operatorname{cosec} \pi s, \tag{11.14}$$

where we have used the product formula for $\sin \pi s$. We can verify eqn (11.10) by showing that the ratio of the two sides is a constant. Equation (11.1) is obtained by evaluation of the limit of

$$\int_0^N t^{\frac{1}{2}s+1}(1-t/N)^N\,\mathrm{d}t \tag{11.15}$$

in two ways as N tends to infinity. We can also obtain from (11.12) the asymptotic formulae

$$\log\Gamma(s) = (s-\tfrac{1}{2})\log s - s + \tfrac{1}{2}\log 2\pi + O(1/|s|), \tag{11.16}$$

and

$$\Gamma'(s)/\Gamma(s) = \log s + O(1/|s|) \tag{11.17}$$

which hold as $|s| \to \infty$ uniformly in any angle $-\pi+\delta < \arg s < \pi-\delta$ for any $\delta > 0$.

Next we consider an L-function $L(s, \chi)$ with χ a proper character $\bmod q$. There are two cases. If $\chi(-1)$ is 1, we argue as above up to

$$\pi^{-\frac{1}{2}s}q^{\frac{1}{2}s}\Gamma(\tfrac{1}{2}s)L(s,\chi) = \int_0^\infty x^{\frac{1}{2}s-1}\sum_{m=1}^\infty \chi(m)\mathrm{e}^{-m^2\pi x/q}\,\mathrm{d}x$$

$$= \tfrac{1}{2}\int_0^\infty x^{\frac{1}{2}s-1}\varphi(x,\chi)\,\mathrm{d}x, \tag{11.18}$$

where

$$\varphi(x,\chi) = \sum_{-\infty}^{\infty} \chi(m)\mathrm{e}^{-m^2\pi x/q}. \tag{11.19}$$

We approach $\varphi(x, \chi)$ through (10.10):

$$\sum_{-\infty}^{\infty} \mathrm{e}^{-(m+\delta)^2\pi/x} = x^{\frac{1}{2}}\sum_{-\infty}^{\infty} \mathrm{e}^{-m^2\pi x}e(m\delta). \tag{11.20}$$

We put $\delta = a/q$ and use eqn (3.8):

$$\chi(m)\tau(\bar{\chi}) = \sum_{a \bmod q}{}^{*}\, \bar{\chi}(a)e_q(am), \tag{11.21}$$

so that

$$\tau(\bar{\chi})\varphi(x,\chi) = \sum_{a \bmod q}{}^{*}\, \bar{\chi}(a)\sum_{m=-\infty}^{\infty} \mathrm{e}^{-m^2\pi x/q}e_q(am)$$

$$= \sum_{a \bmod q}{}^{*}\, \bar{\chi}(a)(q/x)^{\frac{1}{2}}\sum_{m=-\infty}^{\infty} \mathrm{e}^{-(m+a/q)^2\pi q/x}, \tag{11.22}$$

which we may rearrange as

$$(q/x)^{\frac{1}{2}}\sum_{-\infty}^{\infty}\sum_{a \bmod q}{}^{*}\, \bar{\chi}(a)\mathrm{e}^{-(mq+a)^2\pi/xq} = (q/x)^{\frac{1}{2}}\sum_{r=-\infty}^{\infty} \bar{\chi}(r)\mathrm{e}^{-r^2\pi/xq}$$

$$= (q/x)^{\frac{1}{2}}\varphi(1/x,\bar{\chi}). \tag{11.23}$$

This will play the part of the modular relation (10.10). As before, we split up the range of integration in (11.18) and find that

$$\int_0^1 x^{\frac{1}{2}s-1}\varphi(x,\bar{\chi})\,\mathrm{d}x = \int_1^\infty t^{-\frac{1}{2}s-1}(\tau(\bar{\chi}))^{-1}(qt)^{\frac{1}{2}}\varphi(t,\bar{\chi})\,\mathrm{d}x. \tag{11.24}$$

The analogue of (11.7) is now seen to be

$$\pi^{-\frac{1}{2}s}q^{\frac{1}{2}s}\Gamma(\tfrac{1}{2}s)L(s,\chi) = \tfrac{1}{2}\int_1^\infty x^{\frac{1}{2}s-1}\varphi(x,\chi)\,\mathrm{d}x + \frac{1}{2}\frac{q^{\frac{1}{2}}}{\tau(\bar{\chi})}\int_1^\infty x^{-\frac{1}{2}s-\frac{1}{2}}\varphi(x,\bar{\chi})\,\mathrm{d}x. \tag{11.25}$$

As before, the right-hand side of (11.25) converges for all s, so that $L(s,\chi)$ has an analytic continuation over the whole plane, with no singularities. Moreover, $L(s,\chi)$ must have zeros at $0, -2, -4,\ldots$ to cancel the poles of $\Gamma(\frac{1}{2}s)$. We proceed to deduce the functional equation. We have

$$\tau(\bar{\chi}) = \sum_{m \bmod q} \bar{\chi}(m)e_q(m) = \sum_{m \bmod q} \bar{\chi}(-m)e_q(-m)$$
$$= \bar{\tau}(\chi), \tag{11.26}$$

since it was assumed that $\chi(-1) = 1$. By eqn (3.14), since χ is proper $\bmod q$,

$$q^{\frac{1}{2}}/\tau(\bar{\chi}) = \tau(\chi)/q^{\frac{1}{2}}. \tag{11.27}$$

We now see that the right-hand side of (11.25) is $\tau(\chi)q^{-\frac{1}{2}}$ times the corresponding expression with s replaced by $1-s$ and χ by $\bar{\chi}$, which gives the functional equation

$$\pi^{-\frac{1}{2}s}q^{\frac{1}{2}s}\Gamma(\tfrac{1}{2}s)L(s,\chi) = \tau(\chi)q^{-\frac{1}{2}}\pi^{-\frac{1}{2}+\frac{1}{2}s}q^{\frac{1}{2}-\frac{1}{2}s}\Gamma(\tfrac{1}{2}-\tfrac{1}{2}s)L(1-s,\bar{\chi}). \tag{11.28}$$

We now consider characters $\chi(m)$ proper $\bmod q$ with $\chi(-1) = -1$. Since we want to consider a sum from $-\infty$ to ∞, we use $m\chi(m)$ in place of $\chi(m)$. Writing $s+1$ for s in (11.2), we have

$$\pi^{-\frac{1}{2}(s+1)}q^{\frac{1}{2}(s+1)}\Gamma(\tfrac{1}{2}(s+1))L(s,\chi) = \int_0^\infty \sum_{m=1}^\infty m\mathrm{e}^{-m^2\pi x/q}x^{\frac{1}{2}s-\frac{1}{2}}\,\mathrm{d}x$$
$$= \tfrac{1}{2}\int_0^\infty \rho(x,\chi)x^{\frac{1}{2}s-\frac{1}{2}}\,\mathrm{d}x, \tag{11.29}$$

where

$$\rho(x,\chi) = \sum_{m=-\infty}^{\infty} m\chi(m)\mathrm{e}^{-m^2\pi x/q}. \tag{11.30}$$

We find a functional equation for $\rho(x,\chi)$ by differentiating (10.10) with respect to δ. We get

$$\tau(\bar{\chi})\rho(x,\chi) = \mathrm{i}q^{\frac{1}{2}}x^{-\frac{3}{2}}\rho(1/x,\bar{\chi}). \tag{11.31}$$

Arguing as before, we find

$$\pi^{-\frac{1}{2}s-\frac{1}{2}}q^{\frac{1}{2}s+\frac{1}{2}}\Gamma(\tfrac{1}{2}(s+1))L(s,\chi)$$
$$= \tfrac{1}{2}\int_1^\infty \rho(x,\chi)x^{\frac{1}{2}s-\frac{1}{2}}\,\mathrm{d}x + \frac{1}{2}\frac{\mathrm{i}q^{\frac{1}{2}}}{\tau(\bar{\chi})}\int_1^\infty \rho(x,\bar{\chi})x^{-\frac{1}{2}s}\,\mathrm{d}x. \tag{11.32}$$

Again, the integrals on the right of eqn (11.32) converge for all s, so that $L(s,\chi)$ has an analytic continuation; it must have zeros at -1, -3, $-5,\ldots$ to cancel the poles of $\Gamma(\frac{1}{2}(s+1))$, and satisfies the functional equation

$$\pi^{\frac{1}{2}s-1}q^{1-\frac{1}{2}s}\Gamma(1-\tfrac{1}{2}s)L(1-s,\bar{\chi}) = \frac{iq^{\frac{1}{2}}}{\tau(\chi)}\,\pi^{-\frac{1}{2}(s+1)}q^{\frac{1}{2}(s+1)}\Gamma(\tfrac{1}{2}(s+1))L(s,\chi). \qquad (11.33)$$

To check this, we note that when $\chi(-1) = -1$

$$\tau(\bar{\chi}) = -\bar{\tau}(\chi). \qquad (11.34)$$

There is also an analytic continuation of $L(s,\chi)$ when $\chi \bmod q$ is not proper. If χ_1 proper mod f induces $\chi \bmod q$, then for $\sigma > 1$

$$L(s,\chi) = \sum_{m=1}^{\infty}\frac{\chi_1(m)}{m^s}\sum_{\substack{d\mid m\\ d\mid q\\ (d,f)=1}}\mu(d) = \sum_{\substack{d\mid q\\ (d,f)=1}}\frac{\mu(d)\chi_1(d)}{d^s}\sum_{r=1}^{\infty}\frac{\chi_1(r)}{r^s}, \qquad (11.35)$$

when we write $m = dr$. The sum over r in (11.35) is $L(s,\chi_1)$, which has an analytic continuation since $\chi_1 \bmod f$ is proper, and the sum over d is defined for all f. The corresponding functional equation for $L(s,\chi)$ contains the sum over d explicitly. We shall not need this case again.

A number of proofs of the functional equation can be found in Chapter 2 of Titchmarsh (1951).

12

HADAMARD'S PRODUCT FORMULA

Suddenly Christopher Robin began to tell Pooh about some of the things: People called Kings and Queens and something called Factors.

II. 174

In proving the prime-number theorem, Hadamard studied *integral functions of finite order,* that is, functions $f(s)$ regular over the whole plane, with

$$\log|f(s)| \ll |s|^A \tag{12.1}$$

for some constant A, as $|s| \to \infty$. The *order* of $f(s)$ is the lower bound of those A for which an inequality of the form (12.1) holds. Hadamard showed that an integral function of finite order can be written as an infinite product containing a factor $s-\rho$ corresponding to each zero ρ of the function. This generalizes the theorem that a polynomial can be written as a product of linear factors. Weierstrass's definition (11.12) of the gamma function is an example. The product is especially simple when $f(s)$ has order at most unity. The order of $1/\Gamma(s+1)$ is unity, from (11.16). We shall obtain the product formulae for $\xi(s)$ and $\xi(s,\chi)$ given by

$$\xi(s) = s(1-s)\pi^{-\frac{1}{2}s}\Gamma(\tfrac{1}{2}s)\zeta(s) \tag{12.2}$$

and

$$\xi(s,\chi) = (q/\pi)^{\frac{1}{2}(s+a)}\Gamma(\tfrac{1}{2}(s+a))L(s,\chi), \tag{12.3}$$

where χ is proper mod q and $a = 0$ or 1 according to the relation $\chi(-1) = (-1)^a$. Note that (11.9) is just the assertion that $\xi(1-s)$ is equal to $\xi(s)$, and (11.28) or (11.33) implies that

$$|\xi(1-s,\chi)| = |\xi(s,\bar{\chi})|. \tag{12.4}$$

First we show that $\xi(s,\chi)$ has order one. By eqn (5.3), if $\sigma > 0$,

$$L(s,\chi) = \int_1^\infty sx^{-s-1}\sum_{m\leqslant x}\chi(m)\,\mathrm{d}x. \tag{12.5}$$

By Pólya's theorem (4.2), the sum over m is bounded, and thus

$$|L(s,\chi)| \ll q^{\frac{1}{2}}\log q\int_1^\infty |s|x^{-\sigma-1}\,\mathrm{d}x \ll \sigma^{-1}|s|q^{\frac{1}{2}}\log q, \tag{12.6}$$

and for $\sigma \geqslant \frac{1}{2}$ we have

$$\log|L(s,\chi)| \ll \log q + \log|s|. \tag{12.7}$$

Stirling's formula (11.16) is uniform in $\sigma \geqslant \frac{1}{2}$, and we have

$$\log|\xi(s,\chi)| \ll |s|\log q|s| \tag{12.8}$$

uniformly in q and in $\sigma \geqslant \frac{1}{2}$. From (12.4), inequality (12.8) is also true for $\sigma \leqslant \frac{1}{2}$.

To bound $\xi(s)$ we recall eqn (5.14):

$$(1-2^{1-s})\zeta(s) = \sum_{m=1}^{\infty} (-1)^{m-1} m^{-s}, \tag{12.9}$$

valid for $\sigma > 0$. As in the derivation of (12.7) from (12.5), we deduce that the right-hand side of (12.9) is $\ll |s|$ for $\sigma \geqslant \frac{1}{2}$. Now when $\sigma \geqslant \frac{1}{2}$ we have

$$(1-s)/(1-2^{1-s}) \ll |s|, \tag{12.10}$$

and so

$$\log|(1-s)\zeta(s)| \ll \log|s|+1, \tag{12.11}$$

and Stirling's formula gives

$$\log|\xi(s)| \ll |s|(\log|s|+1) \tag{12.12}$$

uniformly in $\sigma \geqslant \frac{1}{2}$. Since $\xi(1-s) = \xi(s)$, (12.12) is also valid for $\sigma \leqslant \frac{1}{2}$. We have now shown that $\xi(s)$ and $\xi(s,\chi)$ have order at most one. If s is real and positive then (12.7) and (12.12) are best possible, by Stirling's formula, so that the order of $\xi(s)$ and $\xi(s,\chi)$ is exactly unity.

We now discuss the zeros ρ of the functions $\xi(s)$ and $\xi(s,\chi)$. First we prove a lemma.

LEMMA. (*Jensen's formula.*) *Let $f(s)$ be a function of the complex variable $s = r\operatorname{cis}\theta$ which is regular in $|s| \leqslant R$ with no zeros on $|s| = R$, for which $f(0) = 1$. Then*

$$(2\pi)^{-1}\int_0^{2\pi} \log|f(R\operatorname{cis}\theta)|\,\mathrm{d}\theta = \int_0^R r^{-1}N(r)\,\mathrm{d}r, \tag{12.13}$$

where $N(r)$ is the number of zeros of $f(s)$ inside the circle $C(r)$: $|s| = r$.

Proof. We can write the left-hand side of (12.13) as

$$(2\pi)^{-1}\int_0^{2\pi}\int_{r=0}^{R} \operatorname{Re}(\mathrm{d}f/f)\,\mathrm{d}r\,\mathrm{d}\theta = (2\pi)^{-1}\int_0^{2\pi}\int_{r=0}^{R} \operatorname{Re}\frac{f'(s)}{f(s)}\operatorname{cis}\theta\,\mathrm{d}r\mathrm{d}\theta$$

$$= \operatorname{Re}\int_{r=0}^{R} \frac{1}{2\pi\mathrm{i}}\int_{\theta=0}^{2\pi} \frac{f'(s)}{f(s)}\,\mathrm{d}s\,\frac{\mathrm{d}r}{r}, \tag{12.14}$$

and the inner integral is $2\pi\mathrm{i}N(r)$.

Now $\xi(0) = 1$, so we can apply (12.13) to $\xi(s)$ at once. We shall prove later that $\xi(0, \chi)$ is non-zero, so that (12.13) can be applied to $\xi(s, \chi)/\xi(0, \chi)$. To avoid a circular argument, we choose a δ for which $\xi(\delta, \chi)$ is non-zero and apply Jensen's formula to $\xi(s, \chi)/\xi(\delta, \chi)$. By (12.7) or (12.12) we have

$$\int_0^R r^{-1}N(r)\,\mathrm{d}r \ll R\log R, \tag{12.15}$$

and as $T \to \infty$

$$N(T) \ll T\log T. \tag{12.16}$$

Here, $N(T)$ is the number of zeros of $\xi(s)$ or of $\xi(s, \chi)$ with $|s| \leqslant T$.

When we examine the formulae (12.2) and (12.3) for $\xi(s)$ and $\xi(s, \chi)$, we see that any zero of $\xi(s, \chi)$ must be a zero of $L(s, \chi)$, and similarly for $\xi(s)$. The converse is not true, because $L(s, \chi)$ has extra zeros at negative integer values to cancel the poles of the gamma function in (12.3). If $s = \sigma + it$ with $\sigma > 1$, Euler's product formula

$$L(s, \chi) = \prod_p \{1 - \chi(p)p^{-s}\}^{-1} \tag{12.17}$$

converges absolutely and so is non-zero. By the functional equation, $\xi(s, \chi)$ is therefore non-zero for $\sigma < 0$, since $\xi(s, \bar{\chi})$ is non-zero for $\sigma > 1$. Thus all zeros $\rho = \beta + \mathrm{i}\gamma$ of $\xi(s, \chi)$ have $0 \leqslant \beta \leqslant 1$, and the same is true for $\xi(s)$ by a similar argument. Riemann's hypothesis is that β is always $\frac{1}{2}$. Riemann stated the hypothesis for $\xi(s)$, but it is difficult to conceive a proof of the hypothesis for $\xi(s)$ that would not generalize to $\xi(s, \chi)$. We shall prove later that $0 < \beta < 1$: this statement is equivalent to the prime-number theorem in the form (5.17) in the sense that each can be derived from the other.

For later use we now prove a result more precise than (12.16).

LEMMA. *The number of zeros $\rho = \beta + i\gamma$ of $\xi(s, \chi)$ in the rectangle B,*

$$0 \leqslant \beta \leqslant 1, \qquad T \leqslant \gamma \leqslant T+1, \tag{12.18}$$

is at most

$$\ll \log(q(|T|+e)), \tag{12.19}$$

and of $\xi(s)$ in B is at most

$$\ll \log(|T|+e). \tag{12.20}$$

Proof. Let s_0 be the point $2 + \mathrm{i}(T+\frac{1}{2})$. Then

$$|L(s_0, \chi)| > 1 - \tfrac{1}{4} - \tfrac{1}{9} - \tfrac{1}{16} - \dots > \tfrac{1}{2}. \tag{12.21}$$

We apply Jensen's formula (12.13) with $R = 3$ and

$$f(s) = \xi(s - s_0, \chi)/\xi(s_0, \chi). \tag{12.22}$$

By (12.7) and Stirling's formula (11.16) the left-hand side of eqn (12.13) is

$$\ll \log q + \log(|T|+e). \tag{12.23}$$

The circle radius 3 and centre s_0 clears the box B by a distance at least $\frac{1}{2}$, so that, if N is the number of zeros required, the right-hand side of eqn (12.13) is

$$\geqslant N \log 6/5. \tag{12.24}$$

This proves (12.19), and (12.20) follows similarly.

We shall not need the following more accurate formula. If $T \geqslant e$, the number of zeros of $\xi(s, \chi)$ with $0 \leqslant \beta \leqslant 1$ and $0 \leqslant \gamma \leqslant T$ is

$$(2\pi)^{-1} T \log(qT/2\pi e) + O(\log qT), \tag{12.25}$$

from which (12.19) and (12.16) follow readily.

We can now prove the product formula. Let $f(s)$ be $\xi(s, \chi)$ or $\xi(s)$, and let ρ run through all zeros of $f(s)$. By (12.19) or (12.20) the series $\sum |\rho|^{-2}$ converges, and so therefore does the product

$$P(s) = \prod_{\rho \neq 0} (1 - s/\rho)\exp(s/\rho), \tag{12.26}$$

where if 0 is a zero we add a factor s at the beginning. $P(s)$ is a regular function with the same zeros as $f(s)$. We should like $f(s)/P(s)$ to be a constant or some other simple function. Certainly

$$g(s) = \log(f(s)/P(s)) \tag{12.27}$$

can be defined to be single-valued and regular over the whole s-plane. We shall prove that

$$g(s) = A + Bs. \tag{12.28}$$

By (12.19) or (12.20), there is a sequence of R tending to infinity with

$$R - |\rho| \gg (\log R)^{-1} \tag{12.29}$$

for each zero ρ. We want a lower bound for $\log P(s)$ on the circle $|s| = R$. Now

$$-\sum_{0 < |\rho| < \frac{1}{2}R} \log|(1 - s/\rho)\exp(s/\rho)| \leqslant R \sum_{0 < |\rho| < \frac{1}{2}R} |\rho|^{-1} \ll R \log^2 R, \tag{12.30}$$

by (12.19) or (12.20), and

$$-\sum_{\frac{1}{2}R \leqslant |\rho| \leqslant 2R} \log|(1 - s/\rho)\exp(s/\rho)| \ll \sum_{\frac{1}{2}R \leqslant |\rho| \leqslant 2R} (1 + \log\log R)$$
$$\ll R \log R \log\log R, \tag{12.31}$$

by (12.29). Finally

$$-\sum_{|\rho| > 2R} \log|(1 - s(\rho)\exp(s/\rho))| \ll R^2 \sum_{|\rho| > 2R} |\rho|^{-2} \ll R \log R. \tag{12.32}$$

On such a circle we have

$$\log|f(s)/P(s)| \ll R \log^2 R. \tag{12.33}$$

The left-hand side of (12.33) is the real part of $g(s)$. If we write

$$s = R \operatorname{cis} \theta,$$

the power-series expansion of $g(s)$ makes $g(R\,\mathrm{cis}\,\theta)$ a Fourier series in θ:

$$g(R\,\mathrm{cis}\,\theta) = \big(a(n)+ib(n)\big)R^n\,\mathrm{cis}\,n\theta, \tag{12.34}$$

so that

$$\mathrm{Re}\,g(R\,\mathrm{cis}\,\theta) = R^n\big(a(n)\cos n\theta - b(n)\sin n\theta\big) \tag{12.35}$$

and

$$\tfrac{1}{2}\pi a(n)R^n = \int_0^{2\pi} \cos(-n\theta)\mathrm{Re}\,g(R\,\mathrm{cis}\,\theta)\mid \mathrm{d}\theta. \tag{12.36}$$

Hence

$$\begin{aligned} |a(n)|R^n &\leqslant \int_0^{2\pi} |\mathrm{Re}\,g(R\,\mathrm{cis}\,\theta)\,\mathrm{d}\theta \\ &\leqslant \int_0^{2\pi} \big(|\mathrm{Re}\,g(R\,\mathrm{cis}\,\theta)|+\mathrm{Re}\,g(R\,\mathrm{cis}\,\theta)-a(0)\big)\,\mathrm{d}\theta \\ &\leqslant 1+\int_0^{2\pi} \max\{\mathrm{Re}\,g(R\,\mathrm{cis}\,\theta),0\}\,\mathrm{d}\theta \\ &\leqslant 1+R\log^2 R. \end{aligned} \tag{12.37}$$

Since (12.37) holds for an infinite sequence of R, $a(2)$, $a(3)$,... must be zero, and similarly so must $b(2)$, $b(3)$,..., and we have proved eqn (12.28).

We have therefore

$$\xi(s) = \mathrm{e}^{Bs}\prod_\rho (1-s/\rho)\exp(s/\rho), \tag{12.38}$$

$$\xi(s,\chi) = C(\chi)\mathrm{e}^{B(\chi)s}\prod_\rho (1-s/\rho)\exp(s/\rho), \tag{12.39}$$

with the modification mentioned above if $\rho = 0$ occurs as a zero. These products converge for all s. By taking logs and differentiating, we get

$$\frac{\zeta'(s)}{\zeta(s)} = B-\frac{1}{s-1}+\tfrac{1}{2}\log 2\pi-\frac{1}{2}\frac{\Gamma'(\frac{1}{2}s+1)}{\Gamma(\frac{1}{2}s+1)}+\sum_\rho\left(\frac{1}{s-\rho}+\frac{1}{\rho}\right). \tag{12.40}$$

and

$$\frac{L'(s,\chi)}{L(s,\chi)} = B(\chi)-\tfrac{1}{2}\log\frac{q}{\pi}-\frac{1}{2}\frac{\Gamma'(\frac{1}{2}(s+a))}{\Gamma(\frac{1}{2}(s+a))}+\sum_{\rho\neq 0}\left(\frac{1}{s-\rho}+\frac{1}{\rho}\right), \tag{12.41}$$

where ρ runs over all zeros of $\zeta(s)$ or $L(s,\chi)$ that do not coincide with the gamma-function poles. We can substitute the equation

$$-\frac{1}{2}\frac{\Gamma'(u)}{\Gamma(u)} = \tfrac{1}{2}\gamma+\sum_1^\infty \frac{1}{2}\left(\frac{1}{u+n-1}-\frac{1}{n}\right), \tag{12.42}$$

which follows from the corresponding product formula (11.13) for $\Gamma(s)$, and obtain a sum over all zeros (except $\rho = 0$) of $\zeta(s)$ or $L(s,\chi)$. It can be shown that

$$B = \log 2+\tfrac{1}{2}\log\pi-1-\tfrac{1}{2}\gamma, \tag{12.43}$$

but no simple expression for $C(\chi)$ and $B(\chi)$ is known in general.

13

ZEROS OF $\xi(s)$

You remember how he discovered the North Pole; well, he was so proud of this that he asked Christopher Robin if there were any other Poles such as a Bear of Little Brain might discover.

I. 131

WE saw in the last chapter that $\xi(s)$ and $\xi(s, \chi)$ have no zeros $\rho = \beta + \mathrm{i}\gamma$ with $\beta > 1$. The prime-number theorem corresponds to the fact that no zeros of $\zeta(s)$ (these include the zeros of $\xi(s)$) have $\beta = 1$. All the direct proofs that $\beta < 1$ at a zero are based on the following argument. The function $-\zeta'(s)/\zeta(s)$ has poles of residue 1 at the poles of $\zeta(s)$ and poles of residue -1 at the zeros. Now

$$-\zeta'(s)/\zeta(s) = \frac{\mathrm{d}}{\mathrm{d}s}\sum_p \log(1-p^{-s})^{-1}$$
$$= \sum_p \log p(1-p^{-s})^{-1}$$
$$= \sum_{m=1}^{\infty} \Lambda(m)m^{-s}, \tag{13.1}$$

where $\Lambda(m)$ is the function of eqn (1.31). At $s = 1$, the series diverges to $+\infty$, corresponding to the pole of residue $+1$. Now, if $1+\mathrm{i}\gamma$ were a zero of $\zeta(s)$, we should expect the series in (13.1) to diverge to $-\infty$, and the partial sums to be as large as those for the case $s = 1$, but negative. To achieve these, the numbers $m^{-\mathrm{i}\gamma}$ must be predominantly near -1. The values of $m^{2\mathrm{i}\gamma}$ are therefore predominantly near $+1$, and there is a pole at $1+2\mathrm{i}\gamma$ with residue $+1$, and so a simple pole of $\zeta(s)$ at $s = 1+2\mathrm{i}\gamma$, which we know does not occur.

To make this argument rigorous, we use (13.1) with $s = \sigma+\mathrm{i}t$ where $\sigma > 1$, so that the series converges. For all real θ,

$$3+4\cos\theta+\cos 2\theta = 2(1+\cos\theta)^2 \geqslant 0. \tag{13.2}$$

Since
$$-\mathrm{Re}(\zeta'(s)/\zeta(s)) = \sum_{m=1}^{\infty} \Lambda(m)m^{-\sigma}\cos(it\log m), \tag{13.3}$$

we have
$$-\frac{3\zeta'(\sigma)}{\zeta(\sigma)} - 4\,\mathrm{Re}\,\frac{\zeta'(\sigma+\mathrm{i}t)}{\zeta(\sigma+\mathrm{i}t)} - \mathrm{Re}\,\frac{\zeta'(\sigma+2\mathrm{i}t)}{\zeta(\sigma+2\mathrm{i}t)} \geqslant 0. \tag{13.4}$$

We now make $\sigma+it$ tend to a zero $\beta+i\gamma$. Since $\zeta(s)$ has a pole at 1, there is a circle centre 1 and some radius r, within which $\zeta(s)$ is non-zero. (Calculation shows that $r = 3$ has this property.) If we suppose that

$$\beta \geqslant 1-\tfrac{3}{5}r, \tag{13.5}$$

then $|\gamma| \geqslant \frac{4}{5}r$, and so is bounded away from zero. In eqn (12.40),

$$\frac{\zeta'(s)}{\zeta(s)} = -B+\frac{1}{s-1}-\tfrac{1}{2}\log 2\pi-\frac{1}{2}\frac{\Gamma'(\frac{1}{2}s+1)}{\Gamma(\frac{1}{2}s+1)}-\sum_{\rho}\left(\frac{1}{s-\rho}+\frac{1}{\rho}\right) \tag{13.6}$$

we shall assume $1 < \sigma \leqslant 2$, $|t| \geqslant \frac{4}{5}r > 0$. Here the sum is over zeros ρ of $\xi(s)$, not over all zeros of $\zeta(s)$, and, since $s-\rho$ and ρ have positive real part, we have

$$\operatorname{Re}\left(\frac{1}{s-\rho}+\frac{1}{\rho}\right) > 0 \tag{13.7}$$

whenever $\sigma > 1$ and $\rho = \beta+i\gamma$ has $0 \leqslant \beta \leqslant 1$. By (11.17) the term in $\Gamma(\frac{1}{2}s+1)$ is $\ll \log(|t|+e)$.

We now write down three inequalities. Since there is a pole at $s = 1$ of residue 1, we have

$$-\zeta'(\sigma)/\zeta(\sigma) = (\sigma-1)^{-1}+O(1). \tag{13.8}$$

At $s = \sigma+i\gamma$ we omit all terms in (13.6) except those from the particular zero with which we are concerned; by (12.7) this gives us the inequality

$$-\zeta'(\sigma+i\gamma)/\zeta(\sigma+i\gamma) \leqslant -(\sigma-\beta)^{-1}+O(\log(|\gamma|+e)). \tag{13.9}$$

Similarly, $$-\zeta'(\sigma+2i\gamma)/\zeta(\sigma+2i\gamma) \leqslant O(\log(|\gamma|+e)). \tag{13.10}$$

When we substitute (13.8), (13.9), and (13.10) into (13.4), we have

$$4(\sigma-\beta)^{-1}-3(\sigma-1)^{-1} \leqslant O(\log(|\gamma|+e)), \tag{13.11}$$

valid as $\sigma \to 1$ from the right. By giving σ a suitable value, we see that

$$\beta < 1-\frac{c}{\log(|\gamma|+e)}, \tag{13.12}$$

where c is an absolute constant. If we constrain c to be less than $\frac{3}{5}r$ then, by (13.5), (13.12) also holds when $|\gamma| \leqslant \frac{4}{5}r$ and is thus valid for all zeros $\rho = \beta+i\gamma$ of $\zeta(s)$.

For twenty years, (13.12) remained the best known upper bound; in 1922 J. E. Littlewood proved that

$$\beta < 1-\frac{c_2\log\log(|\gamma|+e^2)}{\log(|\gamma|+e)}, \tag{13.13}$$

for some constant c_2 (Titchmarsh 1951, theorem 5.17). The latest result is

$$\beta < 1 - \frac{c_3(\epsilon)}{\log(|\gamma|+e)^{\frac{2}{3}+\epsilon}}, \tag{13.14}$$

where the constant c_3 depends on ϵ. This was proved in 1958 by Korobov and by I. M. Vinogradov independently. Intermediate improvements on (13.13) used the intricate methods of I. M. Vinogradov (1954).

14

ZEROS OF $\xi(s, \chi)$

'What do you think you'll answer?'
'I shall have to wait till I catch up with it,' said Winnie-the-Pooh.

I. 34

In this chapter and the next we prove results like (13.12) for the zeros $\rho = \beta + i\gamma$ of $L(s, \chi)$, with uniform constants in the upper bounds. We actually work with $\xi(s, \chi)$, where χ is proper mod q; the zeros of $\xi(s, \chi)$ are those zeros of $L(s, \chi)$ that are not at negative integers and so are not cancelled by gamma-function poles. We use the product formula in the differentiated form (12.41),

$$\frac{L'(s,\chi)}{L(s,\chi)} = B(\chi) - \tfrac{1}{2}\log\frac{q}{\pi} - \frac{1}{2}\frac{\Gamma'(\frac{1}{2}(s+a))}{\Gamma(\frac{1}{2}(s+a))} + \sum_{\rho}\left(\frac{1}{s-\rho} + \frac{1}{\rho}\right), \tag{14.1}$$

where the sum is over zeros ρ of $\xi(s, \chi)$, the term $1/\rho$ being omitted if $\rho = 0$. We shall assume throughout the chapter that $s = \sigma + it$ with $1 < \sigma \leqslant \frac{3}{2}$.

The first complication is the elimination of $B(\chi)$ from eqn (14.1). We subtract from (14.1) its value at $s = 2$, noting that

$$|L'(2,\chi)|/|L(2,\chi)| \leqslant \sum_{m=1}^{\infty} \Lambda(m)m^{-2}, \tag{14.2}$$

which is bounded independently of χ. Estimating the gamma-function term from (11.17), we have

$$-\operatorname{Re}\frac{L'(s,\chi)}{L(s,\chi)} \leqslant -\sum_{\rho}\operatorname{Re}\left(\frac{1}{s-\rho} - \frac{1}{2-\rho}\right) + O(\log(|t|+e)), \tag{14.3}$$

where the term $\rho = 0$, if it occurs, is now included in the sum. We now note that

$$\sum_{\rho}\operatorname{Re}(2-\rho)^{-1} = \sum_{\rho}(2-\beta)|2-\rho|^{-2} \ll \log q, \tag{14.4}$$

by (12.19). Writing

$$l(t) = \log\{q(|t|+e)\}, \tag{14.5}$$

we have

$$-\operatorname{Re}\frac{L'(s,\chi)}{L(s,\chi)} < -\sum_{\rho}\operatorname{Re}\frac{1}{s-\rho} + O(l(t)) \tag{14.6}$$

for $1 < \sigma \leqslant \frac{3}{2}$, the implied constant being absolute.

When we apply (13.2) we obtain the relation

$$-3\frac{L'(\sigma,\chi_0)}{L(\sigma,\chi_0)}-4\,\mathrm{Re}\,\frac{L'(\sigma+it,\chi)}{L(\sigma+it,\chi)}-\mathrm{Re}\,\frac{L'(\sigma+2it,\chi^2)}{L(\sigma+2it,\chi^2)}\geqslant 0, \quad (14.7)$$

valid for $\sigma > 1$, as the analogue of (13.4). Here, χ_0 is the trivial character mod q, whose value $\chi(m)$ is 1 when $(m,q) = 1$ and 0 otherwise, and χ^2 is the character whose value at m is $\{\chi(m)\}^2$. Although we have supposed χ to be proper mod q, χ^2 might be trivial and certainly need not be proper mod q. The trivial character χ_0 is not proper mod q. However, if χ_1 proper mod f induces χ_2 mod q, then $L'(s,\chi_2)/L(s,\chi_2)$ and $L'(s,\chi_1)/L(s,\chi_1)$ differ only by terms involving powers of those primes that divide q but not f. For $\sigma > 1$, these terms give at most

$$\left|\sum_{\substack{m=p^r\\ p|q\\ p\nmid f}}\frac{\chi_1(m)\Lambda(m)}{m^s}\right| \leqslant \sum_{p\leqslant q}\log p\left(\frac{1}{p^\sigma}+\frac{1}{p^{2\sigma}}+\ldots\right)$$

$$\leqslant \sum_{p|q}\log p/(p-1)\leqslant \log q. \quad (14.8)$$

The inequality (14.8) applies also for $f = 1$, $\chi_2 = \chi_0$. We conclude that (14.6) is valid for any non-trivial χ mod q, possibly with a different O-constant, and that

$$-\,\mathrm{Re}\,\frac{L'(s,\chi_0)}{L(s,\chi_0)}\leqslant\frac{\sigma-1}{|s-1|^2}-\mathrm{Re}\sum_\rho\frac{1}{s-\rho}+O(l(t)) \quad (14.9)$$

for the trivial character χ_0 mod q.

If χ^2 is non-trivial, substitution of (14.6) and (14.9) into (14.7) gives

$$4(\sigma-\beta)^{-1}\leqslant 3(\sigma-1)^{-1}+O(l(t)), \quad (14.10)$$

implying that

$$\beta\leqslant 1-c_1/l(\gamma) \quad (14.11)$$

for some absolute constant c_1 when we choose σ appropriately. If χ^2 is trivial, then

$$4(\sigma-\beta)^{-1}\leqslant 3(\sigma-1)^{-1}+(\sigma-1)/\{(\sigma-1)^2+4\gamma^2\}+O\{l(\gamma)\}, \quad (14.12)$$

which is consistent with $\beta = 1$ when $\sigma\to 1$. However, if

$$|\gamma|\geqslant c_2/l(\gamma) \quad (14.13)$$

for some positive c_2, then by choice of σ in (14.12) we can show that

$$\beta\leqslant 1-c_3/l(\gamma), \quad (14.14)$$

with a smaller absolute constant c_3. We have now shown that either (14.14) is true or

$$|\gamma|\leqslant\delta/\log q, \quad (14.15)$$

where δ is an absolute constant. The absolute constant c_3 in (14.14) depends on the choice of δ in (14.15). When (14.15) is satisfied with

$\gamma \neq 0$ we can still deduce an upper bound for β, but it is very close to unity, and tends to 1 as $\gamma \to 0$.

The third and greatest difficulty is to deal with zeros close to unity when χ^2 is trivial. First we show that there is at most one. We have

$$\begin{aligned} -\mathrm{Re}\,\frac{L'(\sigma,\chi)}{L(\sigma,\chi)} &= \mathrm{Re}\sum_1^\infty \frac{\Lambda(m)\chi(m)}{m^\sigma} \\ &\geqslant -\sum_1^\infty \Lambda(m)m^{-\sigma} \\ &\geqslant -(\sigma-1)^{-1}+O(1). \end{aligned} \tag{14.16}$$

If $\rho_1 = \beta_1+i\gamma_1$ and $\rho_2 = \beta_2+i\gamma_2$ are two zeros satisfying (14.15), then

$$\begin{aligned} -\mathrm{Re}\,\frac{L'(\sigma,\chi)}{L(\sigma,\chi)} &\leqslant -\mathrm{Re}\,\frac{1}{\sigma-\rho_1}-\mathrm{Re}\,\frac{1}{\sigma-\rho_2}+O(\log q) \\ &\leqslant -\frac{\sigma-\beta_1}{(\sigma-\beta_1)^2+\gamma_1^2}-\frac{\sigma-\beta_2}{(\sigma-\beta_2)+\gamma_2^2}+O(\log q) \\ &\leqslant \frac{2(\sigma-\beta_1)}{(\sigma-\beta_1)^2+\delta^2(\log q)^{-2}}+O(\log q), \end{aligned} \tag{14.17}$$

if $\beta_2 \geqslant \beta_1 \geqslant 1-\delta/(\log q)$. If δ is small enough, this implies that

$$\beta_1 \leqslant 1-c_4/l(\gamma_1). \tag{14.18}$$

Clearly we can choose $c_4 \leqslant c_3$, and so (14.18) is true for every zero $\beta_1+i\gamma_1$ of $L(s,\chi)$ except (possibly) ρ_2, and the possible exception ρ_2 occurs only if $\chi(m)$ is always real, so that χ^2 is trivial. Since $\bar{\rho}_2$ is also a zero when $L(s,\chi)$ has real coefficients, if ρ_2 fails to satisfy (14.18) we conclude that ρ_2 is real. We devote the next chapter to the case of an *exceptional zero* β on the real axis.

15

THE EXCEPTIONAL ZERO

Piglet said that the best place would be somewhere where a Heffalump was, just before he fell into it, only about a foot farther on.

I. 57

In this chapter we consider *real characters*, that is, characters for which $\chi(m)$ is always real and thus χ^2 is trivial. As far as we know, the corresponding L-functions may have real zeros β with $\frac{1}{2} \leqslant \beta \leqslant 1$. Just as before we saw that $L(s, \chi)$ cannot have two zeros both close to 1, we shall now see that two functions $L(s, \chi)$ corresponding to different proper characters cannot both have zeros close to 1. Suppose χ_1 is proper mod q_1, χ_2 is proper mod q_2, and the corresponding L-functions vanish at β_1 and β_2. In place of (13.2) we use

$$(1+\chi_1(m))(1+\chi_2(m)) \geqslant 0, \tag{15.1}$$

which implies that

$$-\frac{\zeta'(\sigma)}{\zeta(\sigma)}-\frac{L'(\sigma, \chi_1)}{L(\sigma, \chi_1)}-\frac{L'(\sigma, \chi_2)}{L(\sigma, \chi_2)}-\frac{L'(\sigma, \chi_1\chi_2)}{L(\sigma, \chi_1\chi_2)} \geqslant 0, \tag{15.2}$$

where $\chi_1\chi_2$ denotes the character mod q_1q_2 whose value at m is $\chi_1(m)\chi_2(m)$. When χ_1 and χ_2 are different, the character $\chi_1\chi_2$ is non-trivial, and (14.6) gives

$$-L'(\sigma, \chi_1\chi_2)/L(\sigma, \chi_1\chi_2) \leqslant O(\log q_1q_2), \tag{15.3}$$

and for $L(\sigma, \chi_1)$

$$-L'(\sigma, \chi_1)/L(\sigma, \chi_1) \leqslant -(\sigma-\beta_1)^{-1}+O(\log q_1), \tag{15.4}$$

and similarly for χ_2. In place of (14.17) we have

$$(\sigma-\beta_1)^{-1}+(\sigma-\beta_2)^{-1}-(\sigma-1)^{-1} \leqslant O(\log q_1q_2), \tag{15.5}$$

and if $\beta_1 \geqslant \beta_2$ then β_2 at any rate satisfies the relation

$$\beta_2 \leqslant 1-c_1/(\log q_1q_2), \tag{15.6}$$

where c_1 is an absolute constant. We deduce a uniform zero-free region.

By (15.6) and (14.18) there is a constant c_2 with the following property. Let $Q > 1$. Then no L-function formed with a character $\chi \bmod q$ with $q \leqslant Q$ has a zero $\rho = \beta + i\gamma$ with

$$\beta > 1 - c_2/\log\{Q(|\gamma|+e)\} \tag{15.7}$$

except possibly at a point β_1 on the real axis, where $L(s, \chi)$ has at worst a simple zero. All $\chi \bmod q$ with $q \leqslant Q$ for which $L(\beta_1, \chi) = 0$ are induced by the same real character.

To prove the prime-number theorem for an arithmetic progression with common difference q, we need to know that neither $\zeta(s)$ nor any L-function formed with a character $\chi \bmod q$ has a zero $\rho = \beta + i\gamma$ with $\beta = 1$. The proof is simpler if we have β explicitly bounded away from unity. We have to deal only with the case χ real, ρ real. One method is to interpret $L(1, \chi)$ as the density of ideals in a quadratic number field. This gives a very weak bound. We shall prove Siegel's theorem.

THEOREM. *For each $\epsilon > 0$ there is a constant $c(\epsilon)$ such that if*

$$L(\beta_1, \chi) = 0,$$

where χ is a character mod q *then*

$$\beta_1 \leqslant 1 - c(\epsilon) q^{-\epsilon}. \tag{15.8}$$

The constant $c(\epsilon)$ in Siegel's theorem is *ineffective*; that is, the proof does not enable us to calculate it. All previous constants in upper bounds, such as c_2 in (15.7), have been ones we could calculate, given a table of values of $\zeta(s)$ for $|s| \leqslant 3$ and standard inequalities such as Stirling's formula.

Following Estermann's account (1948) of Siegel's theorem, we consider the function

$$F(s) = \zeta(s) L(s, \chi_1) L(s, \chi_2) L(s, \chi_1 \chi_2), \tag{15.9}$$

where χ_1 and χ_2 are real characters proper mod q_1 and mod q_2 respectively. By (15.1), the Dirichlet series

$$\log F(s) = \sum_{m=1}^{\infty} \{1 + \chi_1(m) + \chi_2(m) + \chi_1 \chi_2(m)\} \Lambda(m) m^{-s} \tag{15.10}$$

has non-negative coefficients; it converges for $\sigma > 1$. For $\sigma > 1$ we can take the exponential of (15.10):

$$F(s) = \sum_{m=1}^{\infty} a(m) m^{-s}, \tag{15.11}$$

with non-negative coefficients. $F(s)$ has (at worst) a simple pole at $s = 1$ of residue

$$\lambda = L(1, \chi_1) L(1, \chi_2) L(1, \chi_1 \chi_2), \tag{15.12}$$

and no pole if $\lambda = 0$. Moreover, $F(s)$ has a power-series expansion,

$$F(s) = \sum_{r=0}^{\infty} b(r)(2-s)^r, \tag{15.13}$$

where
$$b(r) = (-1)^r F^{(r)}(2)/r!$$

$$= \frac{(-1)^r}{r!} \sum_{m=1}^{\infty} (-\log m)^r \frac{a(m)}{m^2} \geqslant 0. \tag{15.14}$$

In particular, $b(0) = F(2)$, which is at least unity. The function

$$F(s)-\lambda/(s-1) = \sum_{0}^{\infty} \{b(r)-\lambda\}(2-s)^r \tag{15.15}$$

has no singularities, and so its power series converges everywhere. If $\lambda = 0$, the series on the right of eqn (15.15) is positive on the negative real axis and represents $F(s)$. Since $F(s)$ is zero at $0, -2, -4, \ldots$, we conclude that $\lambda \neq 0$.

We now have $\beta_1 < 1$. To prove the more precise result (15.9), we use Cauchy's formula, integrating round a circle, centre 2 and radius $\frac{3}{2}$, to obtain the coefficients $b(r)$. On the circle, $\zeta(s)$ is bounded, and for the L-functions we use (12.6),

$$|L(s,\chi)| \ll \sigma^{-1}|s|q^{\frac{1}{2}}\log q, \tag{15.16}$$

which was proved from eqn (5.3) and Pólya's theorem, and is thus true for any non-trivial $\chi \bmod q$, proper or not. We have

$$|b(r)-\lambda| = \left|\frac{1}{2\pi i}\int \frac{F(s)}{(s-2)^{r+1}}\,ds\right|$$

$$\ll (\tfrac{2}{3})^r(q_1^{\frac{1}{2}}\log q_1)(q_2^{\frac{1}{2}}\log q_2)(q_1^{\frac{1}{2}}q_2^{\frac{1}{2}}\log q_1 q_2)$$

$$\ll (\tfrac{2}{3})^r q_1 q_2 \log^3 q_1 q_2, \tag{15.17}$$

the constant implied in the $\ll$ sign being absolute. We restrict ourselves to the range $\frac{9}{10} \leqslant \sigma < 1$ and estimate the tail of the series (15.13) for $F(s)$.

$$\sum_{r=R+1}^{\infty} |b(r)-\lambda|(2-\sigma)^r \ll \sum_{R+1}^{\infty} q_1 q_2 \log^3 q_1 q_2 (\tfrac{2}{3}(2-\sigma))^r$$

$$\ll q_1 q_2 \log^3 q_1 q_2 \sum_{R+1}^{\infty} (\tfrac{11}{15})^r$$

$$\ll q_1 q_2 \log^3 q_1 q_2 (\tfrac{11}{15})^R, \tag{15.18}$$

and thus (taking only the first term in the series for $F(s)$)

$$F(\sigma)-\frac{\lambda}{\sigma-1} \geqslant 1-\lambda\sum_{0}^{R}(2-\sigma)^r - O(q_1 q_2 \log^3 q_1 q_2 (\tfrac{11}{15})^R)$$

$$\geqslant 1-\lambda((2-\sigma)^{R+1}-1)/(1-\sigma) - O(q_1 q_2 \log^3 q_1 q_2 (\tfrac{11}{15})^R). \tag{15.19}$$

We choose R so that the error term in (15.19) is less than $\frac{1}{2}$; this is possible with a choice of R with

$$R = A\log q_1 q_2, \tag{15.20}$$

where A is an absolute constant. Now

$$\begin{aligned} F(\sigma) &> \tfrac{1}{2}-\lambda(2-\sigma)^R/(1-\sigma) \\ &\geqslant \tfrac{1}{2}-\lambda(1-\sigma)^{-1}\exp\{R\log(1+(1-\sigma))\} \\ &> \tfrac{1}{2}-\lambda(1-\sigma)^{-1}\exp\{R(1-\sigma)\} \\ &\geqslant \tfrac{1}{2}-B\lambda(1-\sigma)^{-1}(q_1 q_2)^{A(1-\sigma)}, \end{aligned} \tag{15.21}$$

where B is another absolute constant.

We now choose χ_2 so that $L(s, \chi_2)$ has a real zero β_2 with

$$1-\epsilon/(A+1) < \beta_2 < 1; \tag{15.22}$$

if the choice is impossible then (15.8) certainly holds. Since $f(\beta_2) = 0$, we have

$$\tfrac{1}{2} < B\lambda(1-\beta_2)^{-1}(q_1 q_2)^{A(1-\beta_2)} \ll \lambda q_1^{A(1-\beta_2)}, \tag{15.23}$$

where the constant depends on the choice of χ_2 and so on ϵ. This is a lower bound for the product λ of the three L-functions at 1. Since $L(1, \chi_2)$ is a known constant, we seek an upper bound for $L(1, \chi_1\chi_2)$. From eqn (5.3),

$$L(s, \chi) = \int_{1}^{\infty} sx^{-s-1}\sum_{m\leqslant x}\chi(m)\,\mathrm{d}x. \tag{15.24}$$

Let

$$r = q^{\frac{1}{2}}\log q, \tag{15.25}$$

the upper bound in Pólya's theorem. Then

$$|L(s, \chi)| \leqslant \int_{1}^{r}|s|x^{-\sigma-1}x\,\mathrm{d}x + \int_{r}^{\infty}|s|x^{-\sigma-1}r\,\mathrm{d}x. \tag{15.26}$$

If $\sigma \geqslant 1-(\log q)^{-1}$, the first integral is $\ll |s|\log q$ and the second is $\ll |s|$, and so

$$|L(s, \chi)| \ll |s|\log q. \tag{15.27}$$

When we integrate round a circle radius $\frac{1}{2}(\log q)^{-1}$ to find $L'(s, \chi)$, we have for

$$\sigma \geqslant 1-\tfrac{1}{2}(\log q)^{-1} \tag{15.28}$$

the bound

$$|L'(\sigma, \chi)| \ll \log^2 q. \tag{15.29}$$

Hence, if $L(s, \chi_1)$ has a zero β_1 in the range (15.28),

$$\lambda = L(1, \chi_1)L(1, \chi_2)L(1, \chi_1\chi_2) \ll \log q_1 L(1, \chi_1)$$
$$\ll \log q_1(1-\beta_1)L'(\sigma, \chi_1) \quad (15.30)$$

for some σ in $\beta_1 < \sigma < 1$, and so by (15.23) and (15.29) we have

$$1 \ll (1-\beta_1)q_1^{A(1-\beta_2)}\log^3 q_1 \ll (1-\beta_1)q_1^{\epsilon}, \quad (15.31)$$

where the constants implied depend on χ_2 and so on ϵ. If β_1 does not satisfy (15.28), then (15.31) certainly holds, and Siegel's theorem (15.8) follows from (15.31).

16

THE PRIME-NUMBER THEOREM

The clock slithered gently over the mantelpiece, collecting vases on the way.

II. 135

WE can now prove the prime-number theorem in the form (5.17). Combining (5.4) and (5.5), we have

$$\frac{1}{2\pi \mathrm{i}}\int_{\alpha-\mathrm{i}T}^{\alpha+\mathrm{i}T}\frac{u^s}{s}\,\mathrm{d}s=\begin{cases}O(u^\alpha(T|\log u|)^{-1}) & \text{if } u<1,\\ \frac{1}{2}+O(\alpha/T) & \text{if } u=1,\\ 1+O(u^\alpha(T\log u)^{-1}) & \text{if } u>1.\end{cases}\tag{16.1}$$

On the line $\sigma=\alpha$, where $\alpha>1$, the series

$$-\zeta'(s)/\zeta(s)=\sum_{m=1}^{\infty}\Lambda(m)m^{-s}\tag{16.2}$$

converges absolutely. We now suppose that x is an integer plus one-half. Then

$$\frac{1}{2\pi\mathrm{i}}\int_{\alpha-\mathrm{i}T}^{\alpha+\mathrm{i}T}\sum_{m=1}^{\infty}\frac{\Lambda(m)}{m^s}\frac{x^s}{s}\,\mathrm{d}s=\sum_{m<x}\Lambda(m)+O\left(\frac{x^\alpha}{T}\sum_{m=1}^{\infty}\frac{\Lambda(m)}{|\log x/m|}\frac{1}{m^\alpha}\right).\tag{16.3}$$

To estimate the series in the error term of (16.3), we first note that

$$\sum_{m<y}\Lambda(m)\ll y,\tag{16.4}$$

by the sieve upper bound (8.19). Partial summation gives

$$\sum_{m\leqslant\frac{1}{2}x}\frac{\Lambda(m)}{m^\alpha|\log x/m|}+\sum_{m\geqslant 2x}\frac{\Lambda(m)}{x^\alpha\log x/m}\ll\frac{1}{\alpha-1}.\tag{16.5}$$

In the remaining range for m we write $m=x+\frac{1}{2}r$, where r is an odd integer. Thus

$$\begin{aligned}\sum_{\frac{1}{2}x<m<2x}\frac{\Lambda(m)}{x^\alpha\log|x/m|}&\ll x^{-\alpha}\log x\sum_{r=\frac{1}{2}-x}^{2x-2}|\log x(1+\tfrac{1}{2}r/x)|^{-1}\\ &\ll x^{-\alpha}\log x\sum_{r=1}^{2x}x/r\\ &\ll x^{1-\alpha}\log^2 x.\end{aligned}\tag{16.6}$$

We can now write the right-hand side of eqn (16.3) as

$$\psi(x)+O\left(\frac{x^{\alpha}}{T(\alpha-1)}+\frac{x\log^2 x}{T}\right), \tag{16.7}$$

where $\psi(x)$, as usual, is the sum function of the coefficients $\Lambda(m)$ of $-\zeta'(s)/\zeta(s)$.

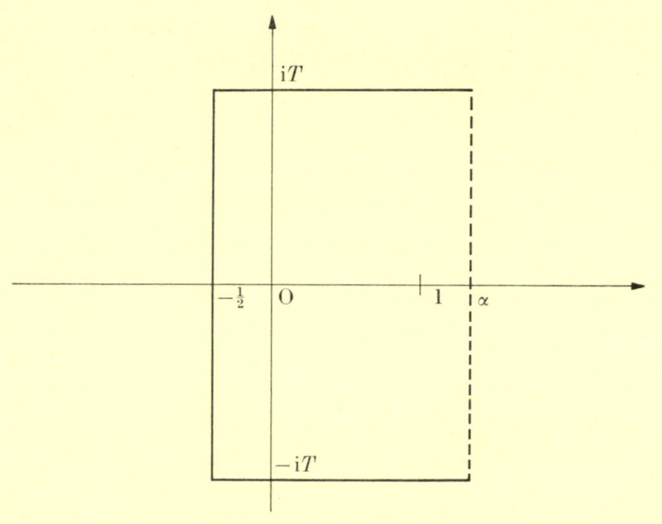

FIG. 2. The contour C

We must now find an approximate value for the integral on the left of eqn (16.3). By (12.20), there is a value of T in each interval

$$t \leqslant T \leqslant t+1$$

for which

$$\big|\,|\gamma|-T\big| \geqslant (\log T)^{-1} \tag{16.8}$$

for each zero $\rho = \beta+\mathrm{i}\gamma$ of $\xi(s)$. With such a value of T we move the integral to the contour C consisting of

C_1: the line segment $[\alpha-\mathrm{i}T, -\frac{1}{2}-\mathrm{i}T]$,

C_2: the line segment $[-\frac{1}{2}-\mathrm{i}T, -\frac{1}{2}+\mathrm{i}T]$,

C_3: the line segment $[-\frac{1}{2}+\mathrm{i}T, \alpha+\mathrm{i}T]$.

In eqn (12.40),

$$\frac{\zeta'(s)}{\zeta(s)} = B-\frac{1}{s-1}+\tfrac{1}{2}\log\pi-\frac{1}{2}\frac{\Gamma'(\frac{1}{2}s+1)}{\Gamma(\frac{1}{2}s+1)}+\sum_{\rho}\left(\frac{1}{s-\rho}+\frac{1}{\rho}\right), \tag{16.9}$$

we subtract the value at $2+\mathrm{i}t$ from that at $\sigma+\mathrm{i}t$, using the fact that

$\zeta'(s)/\zeta(s)$ is bounded on $\sigma = 2$, where its Dirichlet series converges uniformly.

$$\frac{\zeta'(s)}{\zeta(s)} = O(1) - \frac{1}{2}\frac{\Gamma'(\frac{1}{2}(s+1))}{\Gamma(\frac{1}{2}(s+1))} + \frac{1}{2}\frac{\Gamma'(\frac{3}{2}+\frac{1}{2}\mathrm{i}t)}{\Gamma(\frac{3}{2}+\frac{1}{2}\mathrm{i}t)} + \sum_{\rho}\left(\frac{1}{s-\rho} - \frac{1}{2+\mathrm{i}t-\rho}\right)$$

$$= O(\log(|t|+e)) + \sum_{\rho}\frac{2-\sigma}{(\sigma+\mathrm{i}t-\rho)(2+\mathrm{i}-t\rho)}$$

$$\ll \log^2 T, \tag{16.10}$$

where we have used (12.20) and (16.8) on the sum over zeros, and (11.17) on the gamma-function terms.

We can now estimate the integral round C. First

$$\left|\frac{1}{2\pi\mathrm{i}}\int_{C_1}\frac{\zeta'(s)}{\zeta(s)}\frac{x^s}{s}\,\mathrm{d}s\right| \ll \frac{\log^2 T}{T}\int_{-\frac{1}{2}}^{\alpha} x^{\sigma}\,\mathrm{d}\sigma$$

$$\ll x^{\alpha}\log^2 T/(T\log x), \tag{16.11}$$

and similarly for the integral along C_3, whilst

$$\left|\frac{1}{2\pi\mathrm{i}}\int_{C_2}\frac{\zeta'(s)}{\zeta(s)}\frac{x^s}{s}\,\mathrm{d}s\right| \ll x^{-\frac{1}{2}}\log^2 T\int_{C_2}\left|\frac{\mathrm{d}s}{s}\right|$$

$$\ll x^{-\frac{1}{2}}\log^3 T. \tag{16.12}$$

At this stage it is convenient to make the choice of α

$$\alpha = 1+(\log x)^{-1}, \tag{16.13}$$

whereupon we have

$$\left|\frac{1}{2\pi\mathrm{i}}\int_{C}\frac{\zeta'(s)}{\zeta(s)}\frac{x^s}{s}\,\mathrm{d}s\right| \ll \frac{x\log^2 T}{T\log x} + x^{-\frac{1}{2}}\log^3 T. \tag{16.14}$$

The contour has been moved past several poles of the integrand. The residues of $\zeta'(s)/\zeta(s)$ are $+1$ at zeros of $\zeta(s)$, -1 at poles, and so the integral in (16.3) differs from that in (16.14) by

$$x - \sum_{|\gamma|<T}\frac{x^{\rho}}{\rho} - \frac{\zeta'(0)}{\zeta(0)}, \tag{16.15}$$

where the sum is over zeros $\rho = \beta+\mathrm{i}\gamma$ of $\zeta(s)$ with $\beta \geqslant 0$. These are all zeros of $\xi(s)$, and by inequality (13.12) each satisfies the relation

$$\beta \leqslant 1 - c_1/\log(|\gamma|+e). \tag{16.16}$$

By (12.20), the sum over zeros in (16.15) is in modulus

$$\ll x^{1-\delta} \sum_{r \leqslant 2\log T} \frac{1}{2^r} \sum_{|\gamma| \leqslant 2^r} 1$$

$$\ll x^{1-\delta} \log^2 T, \tag{16.17}$$

where

$$\delta = c_1/\log(T+e). \tag{16.18}$$

We have now shown that

$$\psi(x) = x + O\left(\frac{x\log^2 x}{T} + x^{1-\delta}\log^2 T\right), \tag{16.19}$$

and when we choose

$$\log T = (\log x)^{\frac{1}{2}} + O(1), \tag{16.20}$$

where the $O(1)$ is to allow us to find a T for which (16.8) holds, the error terms in (16.19) are bounded by expressions of the form

$$\ll x \exp\{-c(\log x)^{\frac{1}{2}}\} \tag{16.21}$$

with different values of c. Hence there exists an absolute constant c_2 such that

$$\psi(x) = x + O\{x \exp(-c_2(\log x)^{\frac{1}{2}})\}. \tag{16.22}$$

We can quickly deduce from (16.22) that

$$\sum_{p<x} \log p = x + O\{x\exp(-c_3(\log x)^{\frac{1}{2}})\}, \tag{16.23}$$

and integration by parts gives

$$\pi(x) = \sum_{p<x} 1 = \int_2^x \frac{du}{u} + O\{x\exp(-c_4(\log x)^{\frac{1}{2}})\}, \tag{16.24}$$

the classical form of the prime-number theorem. Clearly, only the O-constants in (16.22), (16.23), and (16.24) are affected when we drop the restriction that x be an integer plus one-half.

We have approached the prime-number theorem by the classical route of Riemann's functional equation and Hadamard's product formula. Neither of these is necessary for the proof of eqn (16.22). There are *elementary* proofs of eqn (5.17), ones which do not use the complex variable at all (see for example Hardy and Wright 1960, Chapter 22). The elementary proofs are disappointing in that much extra effort is needed to prove the prime-number theorem with an explicit error term, rather than as an asymptotic equality. An analytic proof that does not use the functional equation is given by Titchmarsh (1951, Chapter 3).

17

THE PRIME-NUMBER THEOREM FOR AN ARITHMETIC PROGRESSION

LET
$$\psi(x,\chi) = \sum_{m\leqslant x} \Lambda(m)\chi(m). \tag{17.1}$$

When χ is proper $\bmod q$, we apply the method of the last chapter to obtain
$$\psi(x,\chi) = O\left(\frac{x\log^2 x}{T}\right) - \sum_{\substack{|\gamma|<T\\ |\rho|>R}} \frac{x^\rho}{\rho} + \frac{1}{2\pi i}\int_C -\frac{L'(s,\chi)}{L(s,\chi)}\frac{x^s}{s}\,ds, \tag{17.2}$$

where α is given by (16.13), and T and R are chosen so that
$$\big||\gamma|-T\big| \gg (\log qT)^{-1}, \tag{17.3}$$
$$\big||\rho|-R\big| \gg (\log q)^{-1} \tag{17.4}$$

for each zero $\rho = \beta+i\gamma$ of $\xi(s,\chi)$. By (12.19), we can suppose that $\frac{1}{5} \leqslant R \leqslant \frac{1}{4}$. The contour C consists of

C_1: the line segment $[\alpha-iT, -\frac{1}{2}-iT]$,

C_2: the line segment $[-\frac{1}{2}-iT, -\frac{1}{2}+iT]$,

C_3: the line segment $[-\frac{1}{2}+iT, \alpha+iT]$,

C_4: the circle centre O, radius R described negatively.

The circle C_4 avoids the pole of x^s/s and a possible L-function zero at $s = 0$.

As in (16.10), the inequality
$$|L'(s,\chi)/L(s,\chi)| \ll \log^2 qT \tag{17.5}$$

holds on the contour C, the constant being absolute. Thus
$$\left|\frac{1}{2\pi i}\int_C \frac{L'(s,\chi)}{L(s,\chi)}\frac{x^s}{s}\,ds\right| \ll \frac{x\log^2 qT}{T\log x} + x^{\frac{1}{4}}\log^2 qT. \tag{17.6}$$

We now treat the sum in (17.2). By (14.18), non-exceptional zeros satisfy the relation
$$\beta \leqslant 1-c_1/\log qT \tag{17.7}$$

with an absolute constant c_1. If q satisfies an inequality
$$\log q \leqslant c_2(\log x)^{\frac{1}{2}}, \tag{17.8}$$

with $c_2 < 1$, we can choose T so that (17.3) holds, and

$$\log qT = (\log x)^{\frac{1}{2}} + O(1). \tag{17.9}$$

As in the last chapter, we deduce

$$\psi(x,\chi) = \begin{cases} -x^{\beta_1}/\beta_1 + O\{x\exp(-c_3(\log x)^{\frac{1}{2}})\} & \text{if there is an exceptional zero } \beta_1, \\ O\{x\exp(-c_3(\log x)^{\frac{1}{2}})\} & \text{if not.} \end{cases} \tag{17.10}$$

The value of c_3 depends on that of c_2 in (17.8), but can be effectively calculated.

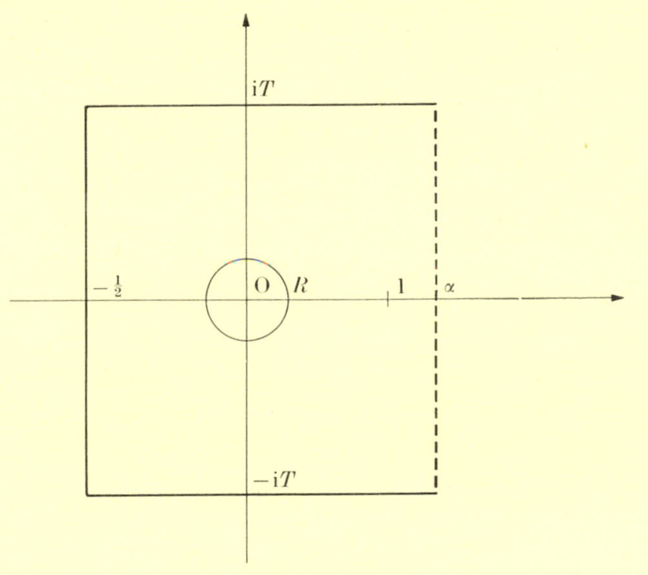

FIG. 3. The contour C

We can absorb the term x^{β_1}/β_1 only if q satisfies

$$q \leqslant \log^N x \tag{17.11}$$

for some $N > 0$. By Siegel's theorem with $\epsilon = \frac{1}{2}N^{-1}$,

$$\beta_1 \geqslant 1 - c_4 q^{-\epsilon} \geqslant 1 - c_4(\log x)^{-\frac{1}{2}}, \tag{17.12}$$

and

$$x^{\beta_1}/\beta_1 \ll x\exp\{-c_4(\log x)^{\frac{1}{2}}\}. \tag{17.13}$$

We have now shown that the inequality

$$|\psi(x,\chi)| \ll x\exp\{-c_5(\log x)^{\frac{1}{2}}\}, \tag{17.14}$$

where c_5 is an absolute constant, holds if (17.11) does, or if (17.8) holds and there is no exceptional zero.

Our characters can at last abandon propriety. If $\chi \bmod q$ is induced by χ_1 proper $\bmod f$ then

$$\psi(x,\chi_1)-\psi(x,\chi) = \sum_{p|q} \log p \sum_{p^r \leqslant x} \chi_1(p^r), \tag{17.15}$$

an expression whose modulus does not exceed

$$\sum_{p|q} \log p \left(1+\frac{\log x}{\log p}\right) \ll \log x \log q. \tag{17.16}$$

The estimate (17.16) also holds if χ is trivial $\bmod q$, $\psi(x,\chi_1)$ being $\psi(x)$. This error term absorbs easily into that of (17.14), which thus holds whenever χ is non-trivial $\bmod q$. For the trivial character $\chi_0 \bmod q$, (17.16) gives

$$|\psi(x,\chi_0)-x| \ll x\exp(-c_6(\log x)^{\frac{1}{2}})+\log x \sum_{p|q} 1, \tag{17.17}$$

and again the first error term will absorb the second.

Let

$$\psi(x;q,a) = \sum_{\substack{m \leqslant x \\ m \equiv a (\bmod q)}} \Lambda(m). \tag{17.18}$$

There are two cases. If (a,q) is not unity, only the powers of primes that divide q can occur in the sum (17.18). If, however, $(a,q) = 1$, we have from eqn (3.3)

$$\psi(x;q,a) = \sum_{\chi \bmod q} \frac{\bar{\chi}(a)}{\varphi(q)} \psi(x,\chi) = \frac{x}{\varphi(q)}+O\{x\exp(-c_7(\log x)^{\frac{1}{2}})\}, \tag{17.19}$$

the prime-number theorem for the arithmetic progression with first term a and common difference q. We have proved (17.19) if (17.8) holds and no character $\bmod q$ has an exceptional zero in its L-function, or if the stronger condition (17.11) on q holds. The result (17.19) subject only to (17.11) is called the Siegel–Walfisch theorem. When $(a,q) = 1$ the arithmetic progression a, $a+q$,... contains infinitely many primes, but we have not proved that the first x/q terms include a prime unless (17.11) holds, so that x is very much larger than q. For a given q we may have to choose x larger still to ensure that the second term in (17.19) is smaller than the first, and $\psi(x;q,a)$ is non-zero. Thus (17.19) is useless for numerical work, since we do not know the value of c_7. From the proof of Siegel's theorem it follows that, if we knew one L-function with an exceptional zero, we should be able to find the c_7 corresponding to one particular value of N in (17.11). If no L-function has an exceptional zero, then (17.19) holds subject only to (17.8), and we can find c_7.

PART III

The Necessary Tools

18

A SURVEY OF SIEVES

> It wasn't what Christopher Robin expected, and the more he looked at it, the more he thought what a Brave and Clever Bear Pooh was.
>
> I. 140

SUPPOSE that we have a classification of the integers into certain classes, and a system of functions σ from the classes to the complex numbers with the properties (i) and (ii) below, which generalize those that the functions

$$\sigma(m) = e_q(am) \tag{18.1}$$

possess for the classification of integers m into residue classes mod q.

(i) Apart from the trivial function σ for which $\sigma(m)$ is always 1, the sum (or integral as appropriate) of $\sigma(m)$ over a complete set of classes is zero.

(ii) If m and n are in different classes there is at least one function for which $\sigma(m) \neq \sigma(n)$.

Then a sequence m_i of integers is uniformly distributed among the classes if and only if

$$\sum_i \sigma(m_i) = o\Big(\sum_i |\sigma(m_i)|\Big) \tag{18.2}$$

for each σ except the trivial one. This is known as Weyl's criterion for uniform distribution. As an example, the prime numbers form a sequence that is uniformly distributed among those residue classes a mod q for which $(a, q) = 1$, in the sense that a proportion $\big(1/\varphi(q)+o(1)\big)$ of the primes up to some bound x falls into each reduced residue class. We deduced this from the fact that $\psi(x, \chi)$ is of smaller order than x when χ is a non-trivial character mod q.

Franel's theorem of Chapter 9 fits this scheme. We were concerned there with the distribution of the F Farey fractions in the interval $[0, 1]$. The appropriate functions $\sigma(m)$ are $e(bf_m)$, where b is an integer, and a calculation with Ramanujan sums shows that

$$\sum_{m=1}^{F} e(bf_m) = \sum_{d|b} dM(Q/d), \tag{18.3}$$

in the notation of Chapter 9. Franel's theorem relates a measure of the uniform distribution of Farey fractions in the interval to the mean square of the sums in eqn (18.3) for varying b.

Many results in number theory rest on the uniform distribution of some sequence among certain classes. This is the final step in the proof of the prime-number theorem for arithmetic progressions. Often the uniformity result is needed at an intermediate stage, and in these cases it is usually sufficient to know (in a quantitative sense) that (18.2) holds most of the time. An assertion that, for a given sequence $m_1, m_2, \ldots$ and set of classes, eqn (18.2) holds for almost all functions σ is called a *large-sieve result*. A common form of large-sieve result is that the mean square of the left-hand side of (18.2) is less than the mean square of the right. The sieve (8.10) of Chapter 8 is such a result with $\sigma(m)$ given by (18.1). In this chapter we shall also consider the characters $\chi(m)$ and the powers m^{-s} as functions $\sigma(m)$. These correspond to distribution among reduced residue classes, and distribution of the sequence within intervals.

The proof of a large-sieve result depends on an upper bound for an average of some convenient auxiliary function. The simplest example is a result for all characters χ to a fixed modulus q. Let $u(m)$ be any complex coefficients and

$$S_\chi = \sum_{m=M+1}^{M+N} u(m)\chi(m). \tag{18.4}$$

In Chapter 7, we introduced the sum

$$S(\alpha) = \sum_{m=M+1}^{M+N} u(m)e(m\alpha); \tag{18.5}$$

and we obtain the result by comparing the identities

$$\sum_{a \bmod q} \left|S\left(\frac{a}{q}\right)\right|^2 = q \sum_{b \bmod q} \left| \sum_{\substack{m=M+1 \\ m \equiv b (\bmod q)}}^{M+N} u(m) \right|^2, \tag{18.6}$$

which follows from eqn (3.1), and

$$\sum_{\chi \bmod q} |S_\chi|^2 = \varphi(q) \sum_{b \bmod q}^{*} \left| \sum_{\substack{m=M+1 \\ m \equiv b (\bmod q)}}^{M+N} u(m) \right|^2, \tag{18.7}$$

which follows from eqn (3.3). The sum on the left of (18.6) can be bounded by the large sieve for exponential sums (7.8), and we have

$$\sum_{\chi \bmod q} |S_\chi|^2 \leqslant q^{-1}\varphi(q)(N+O(q^{-1})) \sum_{m=M+1}^{M+N} |u(m)|^2. \tag{18.8}$$

When we restrict our attention to proper characters we can sum over q on the left of (18.8); by eqn (3.8) for χ proper mod q,

$$\tau(\bar{\chi})\chi(m) = \sum_{a \bmod q}^{*} \bar{\chi}(a) e_q(am), \tag{18.9}$$

and we have

$$\begin{aligned} \tau(\bar{\chi}) S_\chi &= \sum_{m=M+1}^{M+N} \sum_{a \bmod q}^{*} \bar{\chi}(a) u(m) e_q(am) \\ &= \sum_{a \bmod q}^{*} \bar{\chi}(a) S\left(\frac{a}{q}\right). \end{aligned} \tag{18.10}$$

By (3.3) again we have

$$\begin{aligned} \sum_{\chi \bmod q}^{*} \left| \bar{\chi}(a) S\left(\frac{a}{q}\right) \right|^2 &\leqslant \sum_{\chi \bmod q} \left| \bar{\chi}(a) S\left(\frac{a}{q}\right) \right|^2 \\ &\leqslant \varphi(q) \sum_{a \bmod q}^{*} \left| S\left(\frac{a}{q}\right) \right|^2, \end{aligned} \tag{18.11}$$

where the asterisk on the sum over χ indicates a restriction to proper characters. When χ is proper mod q then $|\tau(\bar{\chi})|^2 = q$, and we have

$$\begin{aligned} \sum_{q \leqslant Q} \frac{q}{\varphi(q)} \sum_{\chi \bmod q}^{*} |S_\chi|^2 &\leqslant \sum_{q \leqslant Q} \sum_{a \bmod q}^{*} \left| S\left(\frac{a}{q}\right) \right|^2 \\ &\leqslant (N+O(Q^2)) \sum_{m=M+1}^{M+N} |u(m)|^2 \end{aligned} \tag{18.12}$$

by (8.10), that is, by (7.8) again.

A third sieve result for characters is possible if the coefficients $u(m)$ are 0 whenever m has any factor smaller than Q; for then $(m, q) = 1$, and (18.9) holds for all characters mod q. For example, if $M = 0$, and $u(m)$ is 1 when m is a prime greater than Q and 0 otherwise, we have

$$\begin{aligned} \sum_{q \leqslant Q} \sum_{\chi \bmod q} & \frac{|\tau(\bar{\chi})|^2}{\varphi(q)} |S_\chi|^2 \\ &\leqslant (N+O(Q^2)) \sum_{1}^{N} |u(m)|^2 \\ &\leqslant (N+O(Q^2))(\pi(N)+O(Q)). \end{aligned} \tag{18.13}$$

The left-hand side of (18.13) can be expanded as

$$\sum_{f\leqslant Q} f \sum_{\chi \bmod f}^{*} |S_\chi|^2 \sum_{\substack{q\equiv 0 (\bmod f)\\ q\leqslant Q}} \frac{\mu^2(q/f)}{\varphi(q)}. \tag{18.14}$$

By eqn (8.16), the term $f = 1$ gives

$$(\pi(N))^2 \log Q(1+o(1)), \tag{18.15}$$

which is approximately half the right-hand side of (18.13) when Q is almost $N^{\frac{1}{2}}$. If for some χ $L(s,\chi)$ has an exceptional zero in the sense of (15.7), the term in $|S_\chi|^2$ can be as big as (18.15). This give sanother proof that 'at most one χ has an exceptional zero', but the range for q is not as large as in (15.7) when we work out the details.

To obtain a sieve result for $\sigma(m) = m^{-\sigma}$ we return to first principles and use what we shall refer to as Gallagher's first lemma (1967).

LEMMA. *Let $f(x)$ be a differentiable function of a real variable x. Suppose that $x_1,\ldots, x_R$ are at least δ apart and*

$$X_1+\tfrac{1}{2}\delta \leqslant x_r \leqslant X_2-\tfrac{1}{2}\delta \tag{18.16}$$

for $r = 1,\ldots, R$. Then

$$\sum_{r=1}^{R} |f(x_r)|^2 \leqslant \frac{1}{\delta}\int_{X_1}^{X_2} |f(x)|^2\,\mathrm{d}x + \left(\int_{X_1}^{X_2} |f(x)|^2\,\mathrm{d}x\right)^{\frac{1}{2}} \left(\int_{X_1}^{X_2} |f'(x)|^2\,\mathrm{d}x\right)^{\frac{1}{2}}. \tag{18.17}$$

Proof. We integrate the identity

$$|f(x_r)|^2 = |f(x)|^2 - \int_{x_r}^{x} \frac{\mathrm{d}}{\mathrm{d}x} |f(x)|^2\,\mathrm{d}x \tag{18.18}$$

over an interval of length δ and obtain

$$\begin{aligned}\delta|f(x_r)|^2 &\leqslant \int_{x_r-\frac{1}{2}\delta}^{x_r+\frac{1}{2}\delta} |f(x)|^2\,\mathrm{d}x + \int_{x_r-\frac{1}{2}\delta}^{x_r+\frac{1}{2}\delta} \left|\int_{x_r}^{y} \frac{\mathrm{d}}{\mathrm{d}x}|f(x)|^2\,\mathrm{d}x\right| \mathrm{d}y \\ &\leqslant \int_{x_r-\frac{1}{2}\delta}^{x_r+\frac{1}{2}\delta} |f(x)|^2\,\mathrm{d}x + \int_{x_r-\frac{1}{2}\delta}^{x_r+\frac{1}{2}\delta} \tfrac{1}{2}\delta \left|\frac{\mathrm{d}}{\mathrm{d}x}|f(x)|^2\right| \mathrm{d}x.\end{aligned} \tag{18.19}$$

By (18.16), these intervals are disjoint and subintervals of $[X_1, X_2]$, and

$$\delta \sum_{r=1}^{R} |f(x_r)|^2 \leqslant \int_{X_1}^{X_2} |f(x)|^2\,\mathrm{d}x + \tfrac{1}{2}\delta \int_{X_1}^{X_2} |2f(x)f'(x)|\,\mathrm{d}x, \tag{18.20}$$

and (18.17) follows from Cauchy's inequality.

To obtain large-sieve results from Gallagher's lemma, we arrange the sum on the left of eqn (18.2) to be $f(x)$, different x_r giving the different functions σ. For instance, for $S(\alpha)$ given by (18.5) and $\sigma(m)$ by (18.1), the points x_r are the rationals a/q in their lowest terms, and Gallagher's lemma gives

$$\sum_{q \leqslant Q} \sum_{a \bmod q}^{*} \left| S\left(\frac{a}{q}\right) \right|^2 \leqslant (Q^2 + \pi N) \sum_{m=M+1}^{M+N} |u(m)|^2. \tag{18.21}$$

Like the relation (7.8), (18.21) is an approach to the conjecture (7.7). Whereas (7.8) gives N the conjectured coefficient unity, (18.21) obtains the conjectured coefficient of Q^2 but not that of N.

The application of Gallagher's first lemma to m^{-s} was made by Davenport. Let

$$f(t) = \sum_{m=1}^{N} u(m) m^{-\alpha - \mathrm{i}t}, \tag{18.22}$$

where α is fixed. The first integral on the right of (18.22) is

$$\int_{T_1}^{T_2} |f(t)|^2 \, \mathrm{d}t = \int_{T_1}^{T_2} \sum_m \frac{u(m)}{m^{\alpha+\mathrm{i}t}} \sum_n \frac{\bar{u}(n)}{n^{\alpha-\mathrm{i}t}} \, \mathrm{d}t$$

$$= \sum_m \frac{|u(m)|^2}{m^{2\alpha}} \int_{T_1}^{T_2} \mathrm{d}t + \sum_m \sum_{\substack{n \\ m \neq n}} \frac{u(m)\bar{u}(n)}{m^{\alpha} n^{\alpha}} \int_{T_1}^{T_2} \left(\frac{n}{m}\right)^{\mathrm{i}t} \mathrm{d}t. \tag{18.23}$$

The second term in (18.23) is

$$\sum_m \sum_{\substack{n \\ m \neq n}} \frac{u(m)\bar{u}(n)}{m^{\alpha} n^{\alpha}} \, \frac{(n/m)^{\mathrm{i}T_2} - (n/m)^{\mathrm{i}T_1}}{\mathrm{i} \log(n/m)}$$

$$\leqslant \sum_{m=1}^{N} \sum_{\frac{1}{2}m \leqslant n \leqslant 2m} \frac{2}{|(n-m)/m|} \left(\frac{1}{2} \frac{|u(m)|^2}{m^{2\alpha}} + \frac{1}{2} \frac{|u(n)|^2}{n^{2\alpha}}\right) +$$

$$+ \sum_{m=1}^{N} \left(\sum_{n > 2m} + \sum_{n < \frac{1}{2}m}\right) \frac{1}{\log 2} \left(\frac{1}{2} \frac{|u(m)|^2}{m^{2\alpha}} + \frac{1}{2} \frac{|u(n)|^2}{n^{2\alpha}}\right), \tag{18.24}$$

where we have used the geometric-mean inequality. The coefficient of $|u(m)|^2 m^{-2\alpha}$ in (18.24) is $O(N \log N)$, and thus we can write the right-hand side of (18.23) as

$$(T_2 - T_1 + O(N \log N)) \sum_{m=1}^{N} |u(m)|^2 m^{-2\alpha}. \tag{18.25}$$

Since

$$f'(t) = \sum_{m=1}^{N} \mathrm{i} \log m \, u(m) m^{-\alpha - \mathrm{i}t}, \tag{18.26}$$

and $\log m \leqslant \log N$, the upper bound for the integral of $|f'(t)|^2$ is at most $\log^2 N$ times that for (18.23). We have now proved that if $t_1,\ldots, t_R$ satisfy the relations

$$t_{r+1}-t_r \geqslant \delta \tag{18.27}$$

for $r = 1,\ldots, R-1$ and

$$T_1+\tfrac{1}{2}\delta \leqslant t_r \leqslant T_2-\tfrac{1}{2}\delta \tag{18.28}$$

for $r = 1,\ldots, R$, then

$$\sum_{r=1}^{R} |f(t_r)|^2 \leqslant (\delta^{-1}+\log N)(T_2-T_1+O(N\log N)) \sum_{1}^{N} \frac{|u(m)|^2}{m^{2\alpha}}, \tag{18.29}$$

which is Davenport's sieve, first published by Montgomery (1969*a*).

19

THE HYBRID SIEVE

Kanga said to Roo, 'Drink up your milk first, dear, and talk afterwards.' So Roo, who was drinking his milk, tried to say that he could do both at once . . . and had to be patted on the back and dried for quite a long time afterwards.

I. 151

In the last chapter we obtained sieve results in the form of inequalities with two terms on the right-hand side, one of which corresponded to the maximum size of the summands, the other to the mean square of the function times the number of summands. In the series for $L(s, \chi)$ we can sieve either over χ or over s. Montgomery (1969*a*) sieved over both χ and s and obtained a hybrid result, which is better than the one we should obtain by sieving over the one and summing over the other. P. X. Gallagher has found a simple method of obtaining hybrid sieve results, which we describe here.

The function $\hat{f}(t)$ of a real variable t is said to be the Fourier transform of $f(x)$ when

$$\hat{f}(t) = \int_{-\infty}^{\infty} f(x)e(xt)\,\mathrm{d}x. \tag{19.1}$$

Under suitable conditions (these include the continuity of f at x),

$$f(x) = \int_{-\infty}^{\infty} \hat{f}(t)e(-xt)\,\mathrm{d}t. \tag{19.2}$$

Equations (19.1) and (19.2) imply formally that

$$\begin{aligned} \int_{-\infty}^{\infty} |\hat{f}(t)|^2\,\mathrm{d}t &= \int_{-\infty}^{\infty} \hat{f}(t) \int_{-\infty}^{\infty} \bar{f}(x)e(-xt)\,\mathrm{d}x\mathrm{d}t \\ &= \int_{-\infty}^{\infty} \bar{f}(x) \int_{-\infty}^{\infty} \hat{f}(t)e(-xt)\,\mathrm{d}t\mathrm{d}x \\ &= \int_{-\infty}^{\infty} |f(x)|^2\,\mathrm{d}x, \end{aligned} \tag{19.3}$$

Plancherel's identity. We shall use eqns (19.1) and (19.2) with f a linear combination of the values of

$$F(x) = \begin{cases} \delta^{-1} & \text{if } |x| \leqslant \tfrac{1}{2}\delta, \\ 0 & \text{if } |x| > \tfrac{1}{2}\delta. \end{cases} \tag{19.4}$$

Here

$$\hat{F}(t) = \frac{\sin \pi\delta t}{\pi\delta t}, \tag{19.5}$$

and we can check by means of a contour integral that (19.2) holds except at the points of discontinuity $x = \pm\frac{1}{2}\delta$. We can verify from first principles that the interchange of integrations in the proof of (19.3) is valid. We now deduce Gallagher's second lemma.

Lemma. *Let*

$$S(t) = \sum_{\nu} c(\nu)e(\nu t), \tag{19.6}$$

where the sum is over a finite set of real numbers ν. Let

$$D(x) = \sum_{|N-x| \leqslant (4T)^{-1}} c(\nu). \tag{19.7}$$

Then

$$\int_{-T}^{T} |S(t)|^2 \, dt \leqslant \pi^2 T^2 \int_{-\infty}^{\infty} |D(x)|^2 \, dx. \tag{19.8}$$

Proof. Let $\delta = (2T)^{-1}$ in eqn (19.5). Then

$$D(x) = \delta \sum_{\nu} c(\nu)F(x-\nu) \tag{19.9}$$

and

$$\hat{D}(t) = \delta \sum_{\nu} c(\nu)e(\nu t)\hat{F}(t) = \delta S(t)\hat{F}(t). \tag{19.10}$$

By (19.3),

$$\int_{-\infty}^{\infty} |D(x)|^2 \, dx = \delta^2 \int_{-\infty}^{\infty} |S(t)|^2 |\hat{F}(t)|^2 \, dt \geqslant \delta^2 \int_{-T}^{T} |S(t)|^2 |\hat{F}(t)|^2 \, dt, \tag{19.11}$$

and

$$|\hat{F}(t)| \geqslant 2/\pi \tag{19.12}$$

for $|t| \leqslant T$, which gives (19.8).

We apply the lemma to a *Dirichlet polynomial*, that is, a finite Dirichlet series

$$S(s, \chi) = \sum_{m=1}^{N} a(m)\chi(m)m^{-\sigma-it}, \tag{19.13}$$

so that the numbers ν are of the form $-\log m$, and

$$D(-x) = \sum_{\lambda^{-1}e^x}^{\lambda e^x} a(m)\chi(m), \tag{19.14}$$

where

$$\lambda = \exp((2T)^{-1}). \tag{19.15}$$

The large sieve (18.8) gives

$$\sum_{\chi \bmod q} \left| \sum_{\lambda^{-1}e^x}^{\lambda e^x} a(m)\chi(m) \right| \leqslant ((\lambda - \lambda^{-1})e^x + q) \sum_{\lambda^{-1}e^x}^{\lambda e^x} |a(m)|^2$$

$$\leqslant (T^{-1}e^x + q) \sum_{\lambda^{-1}e^x}^{\lambda e^x} |a(m)|^2. \qquad (19.16)$$

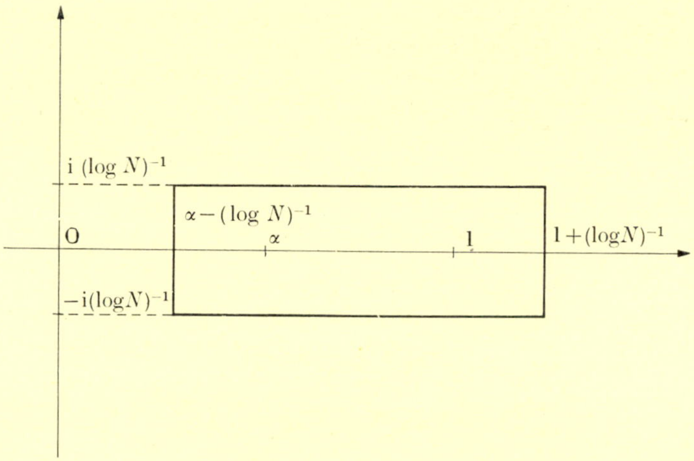

FIG. 4. The contour C

By the lemma,

$$\int_{-T}^{T} \sum_{\chi \bmod q} |S(s,\chi)|^2 \, dt \leqslant T^2 \sum_{m=1}^{N} m^{-2\sigma} \int_{\log(m/\lambda)}^{\log m\lambda} (T^{-1}e^x + q) \, dx$$

$$\leqslant \sum_{m=1}^{N} (m + qT)|a(m)|^2 m^{-2\sigma}. \qquad (19.17)$$

A more realistic application is the following. For each character χ mod q there is a set of points $s(r,\chi) = \sigma(r,\chi) + it(r,\chi)$ within a rectangle

$$\alpha \leqslant \sigma \leqslant 1, \quad -T \leqslant t \leqslant T, \qquad (19.18)$$

satisfying the condition

$$t(r+1,\chi) - t(r,\chi) \geqslant \delta, \qquad (19.19)$$

where δ is a given real number, not necessarily the δ of equation (19.4). Clearly we need Gallagher's first lemma (18.17), but it is not directly applicable, since σ is also a function of r. We turn to Cauchy's identity. Let C be the rectangle (Fig. 4) whose vertices are

$$\alpha - (\log N)^{-1} \pm i(\log N)^{-1}, \quad 1 + (\log N)^{-1} \pm i(\log N)^{-1}.$$

Then if $s = \sigma+it$ satisfies (19.18) we have

$$S^2(s,\chi) = \frac{1}{2\pi i}\int_C \frac{S^2(u+it,\chi)}{u+it-s}\,du, \tag{19.20}$$

and hence
$$|S(s,\chi)|^2 \leqslant \log N \int_C |S(u+it,\chi)|^2\,|du|, \tag{19.21}$$

uniformly in σ. We sum over r and apply (18.17) under the integral sign. Next we sum over $\chi \bmod q$. One of the terms on the right of (18.17) is a geometric mean, but Cauchy's inequality allows us to sum over χ in each factor before taking the geometric mean. We have now to estimate two sums over χ of integrals, which are of the form (19.17). In each, σ and T have been replaced by λ and $T+\frac{1}{2}\delta$, where $u = \lambda+i\tau$, λ and τ real; in the first $a(m)$ has been replaced by $a(m)m^{-i\tau}$, and in the second by $-ia(m)\log m\, m^{-i\tau}$. It is simpler to describe this inequality in words than to write it out! After applying (19.17) we have

$$\sum_{\chi \bmod q}\sum_{r=1}^{R(\chi)} |S(u+it(r,\chi),\chi)|^2 \ll (\delta^{-1}+\log N)\sum_{m=1}^{N}(m+qT)|a(m)|^2 m^{-2\lambda}. \tag{19.22}$$

Finally we integrate round C, noting that

$$\begin{aligned}\int_C m^{-2\lambda}\,|du| &\ll m^{-2\alpha}(\log N)^{-1} + \int_{\alpha-(\log N)^{-1}}^{\alpha+(\log N)^{-1}} m^{-2\lambda}\,d\lambda\\ &\ll m^{-2\alpha}\min\{1-\alpha+(\log N)^{-1}, (\log m)^{-1}\}\\ &\ll \big(m^{2\alpha}\log(m+e)\big)^{-1}.\end{aligned} \tag{19.23}$$

Hence

$$\sum_{\chi \bmod q}\sum_{r=1}^{R(\chi)} |S(s(r,\chi),\chi)|^2 \ll (\delta^{-1}+\log N)\sum_{m=1}^{N}(m+qT)\frac{\log N}{\log(m+e)}\frac{|a(m)|^2}{m^{2\alpha}}, \tag{19.24}$$

where for each χ the points $s_r = s(r,\chi)$ satisfy (19.18) and (19.19).

We can use (18.12) in place of (18.8) and show that

$$\int_{-T}^{T}\sum_{q\leqslant Q}\frac{q}{\varphi(q)}\sum_{\chi \bmod q}^{*}|S(s,\chi)|^2\,dt \leqslant \sum_{m=1}^{N}(m+Q^2T)\frac{|a(m)|^2}{m^{2\sigma}} \tag{19.25}$$

and

$$\begin{aligned}&\sum_{q\leqslant Q}\frac{q}{\varphi(q)}\sum_{\chi \bmod q}^{*}\sum_{r=1}^{R(\chi)}|S(s(r,\chi),\chi)|^2\\ &\qquad\ll (\delta^{-1}+\log N)\sum_{m=1}^{N}(m+Q^2T)\frac{\log N}{\log(m+e)}\frac{|a(m)|^2}{m^{2\alpha}},\end{aligned} \tag{19.26}$$

where the points $s(r,\chi)$ satisfy (19.18) and (19.19) for each χ.

The point of these hybrid sieve results is that we have $N+Q^2T$ (in fact, $m+Q^2T$) in (19.25) etc. in place of $(N+Q^2)T$ or $(N+T)Q^2$ which we would have from sieving one parameter and summing the other. A useful mnemonic for large-sieve results is that $N \sum |a(m)|^2 m^{-2\sigma}$ corresponds to the square of the trivial term $t = 0$, χ trivial, and that there are $O(Q^2T)$ non-trivial terms, whose mean square is $\sum |a(m)|^2 m^{-2\sigma}$. We have certainly shown that eqn (18.2) holds for almost all pairs s, χ.

20

AN APPROXIMATE FUNCTIONAL EQUATION (I)

In a corner of the room, the tablecloth began to wriggle. Then it wrapped itself into a ball and rolled across the room. Then it jumped up and down once or twice, and put out two ears. It rolled across the room again, and unwound itself.

II. 135

We now have efficient tools for averaging finite Dirichlet series. We would like to average powers of $L(s,\chi)$ over points s and characters χ. In this chapter and the next, we express powers of $L(s,\chi)$ as finite sums plus error terms. The best results of this type contain partial sums for $L(s,\chi)$, as one would expect, but also partial sums for $L(1-s,\bar{\chi})$. They have thus earned the nickname of *approximate functional equations*. There are many ways of proving approximate functional equations; we give a straightforward method of H. L. Montgomery's which uses the functional equation explicitly. Our application leads us to consider $L^2(s,\chi)$, where χ is proper mod q. This has the functional equation

$$L^2(1-u,\chi) = (q/\pi)^{2u-1}G(u)L^2(u,\bar{\chi}), \tag{20.1}$$

in which we have put

$$G(u) = G(u,\bar{\chi}) = \frac{q}{\tau^2(\bar{\chi})}\frac{\Gamma^2(\frac{1}{2}(u+a))}{\Gamma^2(\frac{1}{2}(1-u+a))}, \tag{20.2}$$

where a is 0 if $\chi(-1)$ is 1, 1 if $\chi(-1)$ is -1. We shall use u for a complex variable $u = \lambda+i\tau$, where λ and τ are real. By Stirling's formula (11.16), when u tends to infinity in a region $\lambda \ll |\tau|^{\frac{1}{2}}$ we have

$$|\Gamma(u)| = (2\pi)^{\frac{1}{2}}|u|^{\lambda-\frac{1}{2}}e^{-\frac{1}{2}\pi|\tau|}(1+O(|\tau|^{-\frac{1}{2}})), \tag{20.3}$$

and so under the same condition on u,

$$|(q/\pi)^{2u-1}G(u)| \sim (\tfrac{1}{2}q|\tau|/\pi)^{2\lambda-1}. \tag{20.4}$$

We use the ideas of Chapter 5 to get a partial sum for the Dirichlet series of $L^2(s,\chi)$. In the fundamental equation (5.4) the convergence of the integral is only conditional, so we introduce a *kernel* $K(\omega)$ to make the convergence absolute. Let

$$K_1(\omega) = \tfrac{9}{32}\pi^4 e^{\omega}\{(\omega-\tfrac{3}{2}\pi i)(\omega-\tfrac{1}{2}\pi i)(\omega+\tfrac{1}{2}\pi i)(\omega+\tfrac{3}{2}\pi i)\}^{-1} \tag{20.5}$$

and $$K(\omega) = K_1(\omega)+K_1(-\omega) = \tfrac{9}{16}\pi^4 \cosh\omega \prod_{-1}^{2} (\omega-(r-\tfrac{1}{2})\pi i)^{-1}. \tag{20.6}$$

We have normalized so that $K(0)$ is 1. In partial fractions,

$$\frac{K_1(\omega)}{\omega} = \frac{1}{2}\frac{e^{\omega}}{\omega} + \sum_{-1}^{2} \frac{9}{32(2-r)!\,(1+r)!}\,\frac{e^{\omega}}{\omega-(r-\frac{1}{2})\pi i}. \tag{20.7}$$

For real positive x, repeated use of eqn (5.4) gives

$$\frac{1}{2\pi i}\int_{2-i\infty}^{2+i\infty} \frac{x^{\omega}}{m^{\omega}}\frac{K(\omega)}{\omega}\,d\omega = c\left(\frac{m}{x}\right), \tag{20.8}$$

where

$$\left.\begin{array}{l} c(v) = 0 \quad \text{if } v \geqslant e \\ c(v) = \frac{1}{2}-\frac{9}{16}\sin(\frac{1}{2}\pi\log v)-\frac{1}{16}\sin(\frac{3}{2}\pi\log v) \quad \text{if } e^{-1} \leqslant v \leqslant e \\ c(v) = 1 \quad \text{if } 0 < v < e^{-1} \end{array}\right\}. \tag{20.9}$$

Let $s = \sigma+it$ be a fixed complex number with $\frac{1}{4} \leqslant \sigma \leqslant \frac{3}{2}$. The restriction on σ is not essential for the proof, but it is sufficient for our application. Then

$$\sum_{m\leqslant ex} \frac{d(m)\chi(m)}{m^s}\,c\left(\frac{m}{x}\right) = \frac{1}{2\pi i}\int_{2-\sigma-i\infty}^{2-\sigma+i\infty} \frac{x^{\omega}K(\omega)}{\omega}\,L^2(s+\omega,\chi)\,d\omega. \tag{20.10}$$

By construction $K(\omega)$ is an integral function, and the only pole of the integrand in (20.10) is at $\omega = 0$, with residue $L^2(s,\chi)$. The difference between $L^2(s,\chi)$ and the sum in (20.10) is therefore given by an integral along a contour passing to the left of this pole, which we shall take as the vertical line $\operatorname{Re}\omega = -\frac{1}{2}-\sigma$. Equations (20.1) and (20.4) provide bounds for the integrand that justify moving the contour: for example, they imply that

$$|L(-\tfrac{1}{2}+i\tau,\chi)|^2 \ll q^2\tau^2, \tag{20.11}$$

since $L(\frac{3}{2}-i\tau,\bar{\chi})$ is bounded, and that

$$|K(\lambda+i\tau)| \ll \tau^{-4} \tag{20.12}$$

in $-1 \leqslant \lambda \leqslant 2$, $|\tau| \geqslant 10$, so that the integral along $\operatorname{Re}\omega = -\frac{1}{2}-\sigma$ converges absolutely. We change the variable of integration to u

($=\lambda+\mathrm{i}\tau$) where $s+\omega = 1-u$, and use the functional equation (20.1) to write

$$L^2(s,\chi)-\sum_{m\leqslant ex}\frac{d(m)\chi(m)}{m^s}c\left(\frac{m}{x}\right)$$

$$=\frac{1}{2\pi\mathrm{i}}\int\limits_{\frac{3}{2}-\mathrm{i}\infty}^{\frac{3}{2}+\mathrm{i}\infty}\frac{x^{1-s-u}K(s+u-1)}{s+u-1}\left(\frac{q}{\pi}\right)^{2u-1}G(u)L^2(u,\bar\chi)\,du. \quad (20.13)$$

Because of the factor $K(s+u-1)$, the integrand in (20.13) is largest when u is close to $1-s$. Now when u is near $1-s$ the terms (20.4) approximate the $(2\lambda-1)$th power of the constant $\frac{1}{2}q|s|/\pi$. In modulus at any rate the main contribution to (20.13) is an integral of the form (20.10) with s and χ replaced by $1-s$ and $\bar\chi$, and x replaced by y, where the integer y is given by

$$y = [q^2t^2/(4\pi^2x)]. \quad (20.14)$$

More precisely, provided $|\mathrm{Im}(s+u-1)| \leqslant |t|^{\frac{1}{2}}$, we have

$$\frac{1}{q|\tau|}\left|\left(\frac{q}{\pi}\right)^{2u-1}G(u)x^{1-u}y^{1-u}\right| \leqslant \left(\frac{q^2(|t|+|t|^{\frac{1}{2}})^2}{4\pi^2xy}\right)^{\lambda-1}$$

$$\leqslant (1+O(|t|^{-\frac{1}{2}})+O(|y|^{-1}))^{2|t|^{\frac{1}{2}}} \leqslant 1, \quad (20.15)$$

provided that

$$y \geqslant |t|^{\frac{1}{2}}. \quad (20.16)$$

If we choose y by

$$4\pi^2xy = q^2t^2 \quad (20.17)$$

exactly, so that y is not necessarily an integer, the term $O(y^{-1})$ is not present and (20.15) is true in any case; but in this application we shall have $y \geqslant q|t|$ and (20.16) will hold easily.

The series for $L^2(\frac{3}{2}+\mathrm{i}\tau,\bar\chi)$ converges absolutely uniformly, so that we may integrate it term by term. By analogy with the term-by-term integration of (20.10), we break the series up into $\varphi_1+\varphi_3$ or into $\varphi_2+\varphi_4$ (according to whether we take $K_1(s+u-1)$ or $K_1(1-s-u)$), where $\varphi_1,\ldots,\varphi_4$ are given by

$$\left.\begin{aligned}\varphi_1(u) &= \sum_{m<ey} d(m)\bar\chi(m)m^{-u}\\ \varphi_2(u) &= \sum_{m<e^{-1}y} d(m)\bar\chi(m)m^{-u}\\ \varphi_3(u) &= \sum_{m>ey} d(m)\bar\chi(m)m^{-u}\\ \varphi_4(u) &= \sum_{m>e^{-1}y} d(m)\bar\chi(m)m^{-u}\end{aligned}\right\}. \quad (20.18)$$

We write the integral in (20.13) as the sum of four integrals numbered correspondingly $I_1,\ldots, I_4$. First we consider the integral I_3 involving

$\varphi_3(u)K_1(s+u-1)$. When (20.15) is valid, the integrand decreases in modulus as we move the contour to the right. We employ the contour C consisting (in the case $|t| \geqslant 10$) of

C_1: the line segment $(2-i\infty, 2-it-i|t|^{\frac{1}{2}}]$,
C_2: the semicircle centre $2-it$, radius $|t|^{\frac{1}{2}}$ to the right of the line $\lambda = 2$,
C_3: the line segment $[2-it+i|t|^{\frac{1}{2}}, 2+i\infty)$,

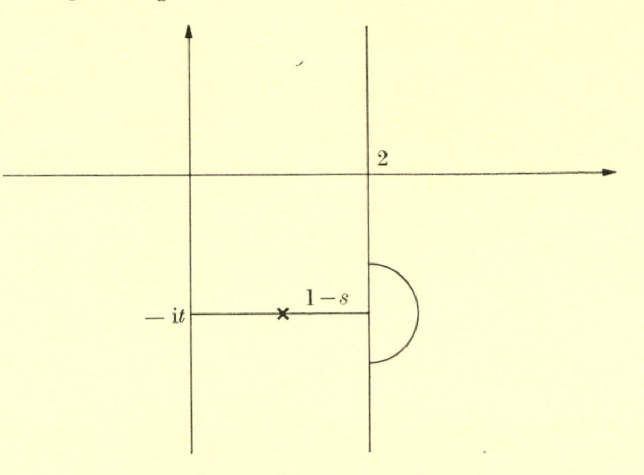

FIG. 5. The contour C

which is the line $\lambda = 2$ with an indentation around the poles of $K_1(s+u-1)$. If $|t| < 10$ we take C to be the line $\lambda = 2$; the estimates below will then hold with $|t|^{-1}$ replaced by 1 and with different implied constants.

For $\lambda \geqslant 2$ we have

$$\begin{aligned}|\varphi_3(u)| &\leqslant \sum_{m>ey} d(m)m^{-\lambda}\\ &= \sum_{m>ey} m^{-\lambda}\{D(m)-D(m-1)\}\\ &\ll (ey)^{1-\lambda}\log y, \end{aligned} \tag{20.19}$$

by partial summation and the estimate (2.13) for the sum function $D(m)$ of the divisor function. By (20.19) and (20.4),

$$\begin{aligned}&\left|\frac{1}{2\pi i}\int_{C_1} \frac{x^{1-s-u}K_1(s+u-1)}{s+u-1}\left(\frac{q}{\pi}\right)^{2u-1} G(u)\varphi_3(u)\,du\right|\\ &\qquad\ll \int_{C_1} \frac{x^{1-\sigma-\lambda}}{(1+|t+\tau|)^5}\left(\frac{q|\tau|}{2\pi}\right)^{2\lambda-1} y^{1-\lambda}\log y\,d\tau\\ &\qquad\ll x^{-\sigma}t^{-2}\left(\frac{xy\pi^2}{4q^2t^2}\right)^{1-\lambda} q|t|\log y\\ &\qquad\ll qx^{-\sigma}|t|^{-1}\log y. \end{aligned} \tag{20.20}$$

The same estimate holds for the integral along C_3. On the semicircle C_2, (20.4) and (20.15) are valid, and we have the upper bound

$$\ll x^{-\sigma}(1+|t|^{\frac{1}{2}})^{-5}q|t|\log y(\pi|t|^{\frac{1}{2}})$$
$$\ll qx^{-\sigma}|t|^{-1}\log y. \qquad (20.21)$$

Hence I_3 and similarly I_4 are bounded by the expression in (20.21).

21

AN APPROXIMATE FUNCTIONAL EQUATION (II)

'Why, what's happened to your tail?' he said in surprise.
'What *has* happened to it?' said Eeyore. 'It isn't there!'
I. 43

WE have now shown that I_3 and I_4 can be regarded as error terms. We treat φ_1 and φ_2 by moving the contour the other way. The first case considered is $|t| > 10$. If $t > 10$ we use the contour D given by

D_1: the line segment $(-\frac{1}{2}-i\infty, -\frac{1}{2}-it-i|t|^{\frac{1}{2}}]$,

D_2: the semicircle, centre $-\frac{1}{2}-it$, radius $|t|^{\frac{1}{2}}$, to the left of $\lambda = -\frac{1}{2}$,

D_3: the line segment $[-\frac{1}{2}-it+i|t|^{\frac{1}{2}}, -\frac{1}{2}-i]$,

D_4: the line segment $[-\frac{1}{2}-i, \frac{1}{4}-i]$,

D_5: the line segment $[\frac{1}{4}-i, \frac{1}{4}+i]$,

D_6: the line segment $[\frac{1}{4}+i, -\frac{1}{2}+i]$,

D_7: the line segment $[-\frac{1}{2}+i, -\frac{1}{2}-i\infty)$.

The contour D for $t < -10$ is the reflection in the real axis of that described above. The indentation about the origin avoids a possible double pole of $G(u)$ at $u = 0$. The analogue of (20.19) is that, uniformly in $\lambda \leqslant \frac{1}{4}$,

$$|\varphi_1(u)| \ll (ey)^{1-\lambda}\log y. \tag{21.1}$$

Let
$$I_1 = \frac{1}{2\pi i}\int_D \frac{x^{1-s-u}K_1(s+u-1)}{s+u-1}\left(\frac{q}{\pi}\right)^{2u-1} G(u)\varphi_1(u)\,du. \tag{21.2}$$

Using (21.1) in place of (20.19) we see that

$$|I_1| \ll qx^{-\sigma}|t|^{-1}\log y; \tag{21.3}$$

and the right-hand side of (21.3) is similarly an upper bound for I_2 in which $K_1(1-s-u)\varphi_2(u)$ replaces $K_1(s+u-1)\varphi_1(u)$.

Between the contours C and D lie the poles of $w^{-1}K(w)$ at $w = 0$ and $w = \pm(r-\frac{1}{2})\pi\mathrm{i}$, where $w = s+u-1$. The residues at these can be calculated term by term; in total they give

$$\sum_{m \leqslant \mathrm{e}y} \frac{d(m)\bar{\chi}(m)}{m^{1-s}} c'\left(\frac{m}{y}\right), \tag{21.4}$$

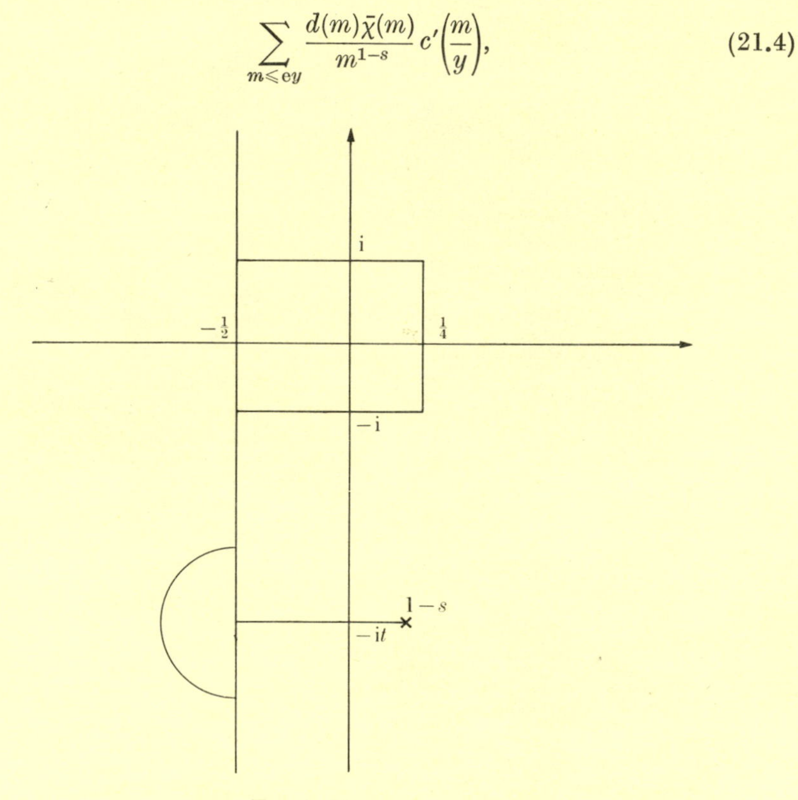

FIG. 6. The contour D

where

$$\left.\begin{aligned} c'(v) &= 0 \quad \text{if } v \geqslant \mathrm{e} \\ c'(v) &= \frac{1}{2}\left(\frac{q}{\pi}\right)^{1-2s} G(1-s) + \\ &\quad + \sum_{-1}^{2} \frac{9}{32} \frac{1}{(1+r)!\,(2-r)!} \left(\frac{q^2\mathrm{e}}{\pi^2 vxy}\right)^{(r-\frac{1}{2})\pi\mathrm{i}} \times \\ &\quad \times G\big(1-s+(r-\tfrac{1}{2})\pi\mathrm{i}\big) \qquad \text{if } \mathrm{e}^{-1} \leqslant v < \mathrm{e} \\ c'(v) &= \left(\frac{q}{\pi}\right)^{1-2s} G(1-s) \quad \text{if } 0 < v < \mathrm{e}^{-1} \end{aligned}\right\} \tag{21.5}$$

If $\chi(-1)$ is -1 then $G(u)$ is regular at 0. For $|t| \leqslant 10$ we take the contour D to be the line $\lambda = -\frac{1}{2}$. The bound for I_1 and I_2 on this contour is the same, but without the factor $|t|^{-1}$ and with a different implied constant. If $\chi(-1)$ is $+1$, however, there is a double pole. If σ is bounded away from 1 we can run the contour D between the origin and the poles of $K_1(s+u-1)/(s+u-1)$. If $\sigma < 1$ we take D as

D_1: the line segment $(-\frac{1}{2}-i\infty, -\frac{1}{2}-i]$,
D_2: the line segment $[-\frac{1}{2}-i, \frac{1}{2}(1-\sigma)-i]$,
D_3: the line segment $[\frac{1}{2}(1-\sigma)-i, \frac{1}{2}(1-\sigma)+i]$,
D_4: the line segment $[\frac{1}{2}(1-\sigma)+i, -\frac{1}{2}+i]$,
D_5: the line segment $[-\frac{1}{2}+i, -\frac{1}{2}+i\infty)$.

The upper bound for the integral along D_3 will be

$$\ll qx^{-\sigma}(1-\sigma)^{-2}\log y. \tag{21.6}$$

If all else fails, we can take D as the vertical line $\lambda = -\frac{1}{2}$ and get an explicit but very complicated extra term in the approximate functional equation from the residue at $u = 0$ of the multiple pole. When $|t| \leqslant 10$, (20.14) is interpreted as

$$q^2 \ll xy \ll q^2. \tag{21.7}$$

We have now shown that an equation of the form

$$L^2(s,\chi) = \sum \frac{d(m)\chi(m)}{m^s}\, c\left(\frac{m}{x}\right) + \sum \frac{d(m)\bar{\chi}(m)}{m^{1-s}}\, \acute{c}\left(\frac{m}{x}\right) + I_1 + I_2 + I_3 + I_4 \tag{21.8}$$

holds both in $\frac{1}{4} \leqslant \sigma \leqslant \frac{5}{4}$, $|t| > 10$, in which case each of $I_1,\ldots, I_4$ is

$$\ll qx^{-\sigma}|t|^{-1}\log y, \tag{21.9}$$

and in $\frac{1}{4} \leqslant \sigma < 1$, $|t| \leqslant 10$, when each of $I_1,\ldots, I_4$ is

$$\ll (1-\sigma)^{-2}qx^{-\sigma}\log y, \tag{21.10}$$

the estimates being uniform in the ranges stated.

Although we have assumed $q > 1$, so that $L(s,\chi)$ is not $\zeta(s)$, we can treat $\zeta^2(s)$ by the same method. We assume $|t| \geqslant 10$, since it would be unprofitable to try to approximate near the double pole at $s = 1$. When we move the integral in (20.13) to the line $\mathrm{Re}\, w = -\frac{1}{2}-\sigma$, this double pole gives a residue

$$2Aw^{-1}K(w) + \frac{d}{dw}\left(w^{-1}K(w)\right) \tag{21.11}$$

evaluated at $w = 1-s$, where A is the second coefficient (in fact, A is Euler's constant γ) in the Laurent expansion

$$\zeta(s) = (s-1)^{-1} + A + O(s-1) \tag{21.12}$$

of $\zeta(s)$ about the pole at $s = 1$. We must add to eqn (21.8) the additional terms (21.11); they are

$$\ll |t|^{-5} \tag{21.13}$$

uniformly in $|t| \geqslant 10$, $\frac{1}{4} \leqslant \sigma \leqslant \frac{5}{4}$.

There are many approximate functional equations in the literature. The one we have proved is easy to sieve, but has more complicated coefficients. Approximate functional equations for $\zeta(s)$ and for $\zeta^2(s)$ are given by Titchmarsh (1951, Chapter 4), and generalized by Chandrasekharan and Narasimhan (1963). A. F. Lavrik has a number of papers on the subject in the *Doklady* and *Izvestia* of the USSR Academy of Sciences. Fogels (1969) has a form rather like that of Montgomery's given here.

22

FOURTH POWERS OF L-FUNCTIONS

They were out of the snow now, but it was very cold, and to keep themselves warm they sang Pooh's song right through six times, Piglet doing the tiddley-poms and Pooh doing the rest of it, and both of them thumping on top of the gate with pieces of stick at the proper places.

II. 7

THE approximate functional equation for $L^2(s, \chi)$ is used to obtain upper bounds for $|L(s, \chi)|$ at individual values of s, or on average. Our inspiration is the Lindelöf hypothesis,

$$|L(\tfrac{1}{2}+\mathrm{i}t, \chi)| \ll (qt)^{\epsilon}, \tag{22.1}$$

for each $\epsilon > 0$, with a constant depending on ϵ. In this chapter, we show that the mean fourth power of $L(s, \chi)$ satisfies the inequality (22.1). Beyond the fourth power, the sums in the approximate functional equation are too long for us to prove (22.1) even on average. We use the hybrid sieve (19.26),

$$\sum_{q \leqslant Q} \frac{q}{\varphi(q)} \sum_{\chi \bmod q}^{*} \sum_{r=1}^{R(\chi)} |S(s(r, \chi), \chi)|^2$$
$$\ll (\delta^{-1}+\log N) \sum_{m=1}^{N} (m+Q^2T) \frac{\log N}{\log(m+\mathrm{e})} \frac{|a(m)|^2}{m^{2\alpha}}. \tag{22.2}$$

Although (22.2) allows us to vary σ, we take $\sigma = \frac{1}{2}$ throughout, since

$$|(q/\pi)^{1-2s}(x/\mathrm{e})^{-r\pi\mathrm{i}}G(1-s+r\pi\mathrm{i})| = 1 \tag{22.3}$$

for $\sigma = \frac{1}{2}$, the gamma functions in G being evaluated at conjugate complex points. The proof is no simpler, but the form of the upper bounds is less complicated.

Cauchy's inequality applied to eqn (21.8) gives

$$|L(s, \chi)|^4 \leqslant 6\left|\sum d(m)\chi(m)m^{-s}c(m/x)\right|^2+$$
$$+6\left|\sum d(m)\bar{\chi}(m)m^{s-1}c'(m/y)\right|^2+6\sum_{r=1}^{4}|I_r|^2, \tag{22.4}$$

and we have a similar result for $\zeta(s)$, with 7 instead of 6 and an extra term $O(|t|^{-10})$. The integers x and y in (22.4) are connected by (20.14):

$$y = [q^2t^2/(4\pi^2 x)]. \tag{22.5}$$

When we fix x and average over χ and t, y is varying. For a good upper bound, x and y must be of the same order of magnitude. We restrict ourselves for the moment to $P < q \leqslant 2P$ and $U < |t| < 2U$, and average the right-hand side of (22.4) over all integer values of x between $\frac{1}{4}PU/\pi$ and $\frac{1}{2}PU/\pi$. For each fixed χ and t the corresponding values of y given by (22.5) are distinct and lie between $\frac{1}{2}PU/\pi - 1$ and $16PU/\pi$. The average of the square of the terms involving y taken over all integers y in this range is at least $\frac{1}{64}$ times the corresponding average over the values of y that actually occur in the sum. This device allows us to sum a Dirichlet series of fixed length $\mathrm{e}y$ over varying χ and t, provided that x does not occur in the coefficients (and vice versa).

For each value of x in the above range, (22.2) gives

$$\sum_{p<q\leqslant 2P} \frac{q}{\varphi(q)} \sum_{\chi \bmod q}{}^{*} \sum_{\substack{r=1 \\ U<|t|\leqslant 2U}}^{R(\chi)} \left| \sum_{m \leqslant \mathrm{e}x} \frac{d(m)\chi(m)}{m^{s(r,\chi)}} c\left(\frac{m}{x}\right) \right|^2$$

$$\ll (\delta^{-1} + \log x) \sum_{m=1}^{\mathrm{e}x} (m + P^2U) \frac{d^2(m) \log x}{m \log(m+\mathrm{e})}$$

$$\ll P^2U \log^5(PU+\mathrm{e}), \tag{22.6}$$

where we have used (2.24) for $\sum d^2(m)/m$, and partial summation. We have assumed

$$\delta \geqslant 1. \tag{22.7}$$

It is only slightly harder to average over the reflected series. For $\mathrm{e}^{-1}y \leqslant m < \mathrm{e}y$ we must break the coefficients $c'(m/y)$ up into five terms corresponding to the five poles of $K_1(w)/w$. We take out a factor

$$\left(\frac{q}{\pi}\right)^{1-2s} \left(\frac{\pi^2 x}{\mathrm{e}q^2}\right)^{-r\pi\mathrm{i}} G(1-s+r\pi\mathrm{i}) \tag{22.8}$$

from the term corresponding to the pole of $K_1(w)/w$ at $r\pi\mathrm{i}$, where $r = 0, \pm\frac{1}{2}, \pm\frac{3}{2}$. By (22.3), the modulus of the expression (22.8) is fixed independently of x, q, t, and the parameter a in the definition of $G(u)$, which is 1 if $\chi(-1) = -1$ and 0 if $\chi(-1) = +1$. After this factor has been removed, a sum like that in (22.6) remains, with no gamma-function factors and no concealed dependence on x. For each r, the right-hand side of (22.6) is an upper estimate for this sum, possibly with a different O-constant.

The contour integrals $I_1, \ldots, I_4$ present complications. First we fix t and average only over characters. We have for any function $F(u)$ for which the integrals exist

$$\left|\int_D F(u)\,\mathrm{d}u\right|^2 \ll \left(\int_D \frac{|\mathrm{d}u|}{|s+u-1|^2}\right)\left(\int_D |F(u)|^2|s+u-1|^2\,|\mathrm{d}u|\right), \tag{22.9}$$

and similarly for integrals along the contour C. The first factor here is $\ll |t|^{-\frac{1}{2}}$. We shall find an estimate for the average of I_1. Here,

$$F(u) = x^{1-s-u}(s+u-1)^{-1}K_1(s+u-1)(q/\pi)^{2u-1}G(u,\chi)\varphi_1(u). \tag{22.10}$$

We wish to average this over χ with the length of the sum $\varphi_1(u)$ fixed, that is, with y fixed. By (22.4), x is approximately varying as q^2, and $(q^2/u)^\lambda$ is bounded on the contour D by a function of t and y alone, by (20.14). Similarly, (20.4) permits us to take out $(q/\pi)^{2u-1}G(u)$ at its maximum. By (18.12),

$$\begin{aligned}\sum_{P<q\leqslant 2P}\frac{q}{\varphi(q)}\sideset{}{^*}\sum_{\chi \bmod q}|\varphi_1(u)|^2 &\ll (y+P^2)\sum_{m\leqslant \mathrm{e}y}\frac{d^2(m)}{m^{2\lambda}}\\ &\ll (y+P^2)(\mathrm{e}y)^{1-2\lambda}\log^3 y,\end{aligned} \tag{22.11}$$

where we have used the estimate

$$\sum_{m\leqslant M} d^2(m) \ll \log^3 M, \tag{22.12}$$

which can be deduced from (2.24). The sum of the integrals involving $F(u)$ on the right-hand side of (22.9) is now

$$\begin{aligned}&\ll \max_{(q,x)}\max_{u\in D}\frac{|t|^{\frac{1}{2}}}{(|t|^{\frac{1}{2}})^8}x^{-2\sigma}\left|\left(\frac{x}{\mathrm{e}}\right)^{2-2u}\left(\frac{q}{\pi}\right)^{4u-2}G^2(u)\right|(y+P^2)(\mathrm{e}y)^{1-2\lambda}\log^3 y\\ &\ll (\max x)^{1-2\sigma}|t|^{-\frac{7}{2}}(y+P^2)\log^3 y\\ &\ll (P^2+P|t|)|t|^{-\frac{7}{2}}\log^3(P(|t|+\mathrm{e})),\end{aligned} \tag{22.13}$$

where we have used (20.15). Hence

$$\sum_{P<q\leqslant 2P}\frac{q}{\varphi(q)}\sideset{}{^*}\sum_{\chi \bmod q}|I_1|^2 \ll (P^2+P|t|)|t|^{-4}\log^3(P(|t|+\mathrm{e})), \tag{22.14}$$

and similarly the same upper bound holds for the corresponding sum involving I_2.

To apply a Gallagher sieve we need a similar bound for a sum involving $\frac{\mathrm{d}}{\mathrm{d}t}I_1$. The contour D varies with t, but this was only to help estimation, and we can keep D fixed and differentiate the integrand. A similar calculation will now give

$$\sum_{P<q\leqslant 2P} \frac{q}{\varphi(q)} \sum_{\chi \bmod q}^{*} \left|\frac{\mathrm{d}}{\mathrm{d}t} I_1\right|^2 \ll (P^2+P|t|)|t|^{-4}\log^5(P(|t|+\mathrm{e})). \tag{22.15}$$

Gallagher's lemma (18.17) gives

$$\sum_{\substack{r=1\\ U<t_r\leqslant 2U}}^{R} |I_1|^2 \leqslant \int_{U-\frac{1}{2}}^{2U+\frac{1}{2}} |I_1|^2\,\mathrm{d}t + \left(\int_{U-\frac{1}{2}}^{2U+\frac{1}{2}} |I_1|^2\,\mathrm{d}t\right)^{\frac{1}{2}} \left(\int_{U-\frac{1}{2}}^{2U+\frac{1}{2}} \left|\frac{\mathrm{d}}{\mathrm{d}t} I_1\right|^2 \mathrm{d}t\right)^{\frac{1}{2}}. \tag{22.16}$$

We sum (22.16) over χ (using Cauchy's inequality on the second term) to find

$$\sum_{P<q\leqslant 2P} \frac{q}{\varphi(q)} \sum_{\chi \bmod q}^{*} \sum_{r=1}^{R(\chi)} |I_1|^2 \ll (P+U)PU^{-3}\log^4(P(U+\mathrm{e})) \tag{22.17}$$

when we substitute in (22.14) and (22.15). The same upper bound holds for the sum with I_2.

The integrals I_3 and I_4 present a further complication in that φ_3 and φ_4 are not finite sums. To treat φ_3 we write

$$\left|\sum_{m>\mathrm{e}y} d(m)\bar{\chi}(m)m^{-u}\right|^2 \leqslant \left(\sum_{n=1}^{\infty} n^{-2}\right)\left(\sum_{n=1}^{\infty} n^2 \left|\sum_{\mathrm{e}^n y<m\leqslant \mathrm{e}^{n+1}y} d(m)\bar{\chi}(m)m^{-u}\right|^2\right). \tag{22.18}$$

By (18.11) we have

$$\sum_{P<q\leqslant 2P} \frac{q}{\varphi(q)} \sum_{\chi \bmod q}^{*} \left|\sum_{\mathrm{e}^n y<m\leqslant \mathrm{e}^{n+1}y} \frac{d(m)\bar{\chi}(m)}{m^u}\right|^2 \ll (\mathrm{e}^{n+1}y+P^2)(\mathrm{e}^n y)^{1-2\lambda}(n+\log y)^3, \tag{22.19}$$

and the sum of terms of the form (22.19) from $n = 1$ to infinity converges uniformly in λ to

$$\ll (y+P^2)(\mathrm{e}y)^{1-2\lambda}\log^3 y. \tag{22.20}$$

When we use (22.20) in place of (22.11) we show that the sums over I_3 and I_4 also satisfy the bound (22.17).

We have now shown that

$$\sum_{P<q\leqslant 2P} \frac{q}{\varphi(q)} \sum_{\chi \bmod q}^{*} \sum_{\substack{r=1 \\ U<t(r,\chi)\leqslant 2U}}^{R(\chi)} |L(\tfrac{1}{2}+it(r,\chi),\chi)|^4$$

$$\ll P^2U\log^5(P(U+e))+(P+U)PU^{-3}\log^4(P(U+e)). \quad (22.21)$$

By (21.13), if the zeta function is included in the sum there is an extra term U^{-9}, which is easily absorbed in the second term on the right of (22.21).

When we sum over values of P and U running through powers of two, we can show that

$$\sum_{q\leqslant Q} \frac{q}{\varphi(q)} \sum_{\chi \bmod q}^{*} \sum_{r=1}^{R(\chi)} |L(\tfrac{1}{2}+it(r,\chi),\chi)|^4 \ll Q^2T\log^5 QT, \quad (22.22)$$

provided that for $r = 1,\ldots, R(\chi)$ we have

$$-T \leqslant t(r,\chi) \leqslant T, \quad (22.23)$$

and for $r = 1,\ldots, R(\chi)-1$ we have

$$t(r+1,\chi)-t(r,\chi) \geqslant 1, \quad (22.24)$$

and for $q = 1$ $|t(r,\chi)| \geqslant 1$.

We could replace (22.24) by a weaker condition

$$t(r+1,\chi)-t(r,\chi) \gg (\log QT)^{-1}, \quad (22.25)$$

and still have the upper bound (22.22), so that the root mean fourth power of $L(s,\chi)$ is $\ll \log QT$. Without the sieving over T this method proves that the root mean fourth power at a fixed s is $\ll \log(Q(|s|+e))$.

A simpler proof of a fourth-power moment is in Gallagher (1967). Another result is given by Linnik (1964, section 41). These relate to summing over χ. The situation is much easier if we merely integrate over t; there are examples in Titchmarsh (1951, Chapter 7). The hybrid moment we have proved here is sketched in Montgomery's paper (1969*b*), but with a larger logarithm power.

PART IV

Zeros and Prime Numbers

23

INGHAM'S THEOREM

> He tried Counting Sheep, which is sometimes a good way of getting to sleep, and, as that was no good, he tried counting Heffalumps.
>
> I. 62

WE shall estimate the number of zeros $\rho = \beta + i\gamma$ of $L(s,\chi)$ or of $\zeta(s)$ in a rectangle

$$\alpha \leqslant \beta \leqslant 1, \qquad -T \leqslant \gamma \leqslant T \tag{23.1}$$

where $\alpha > \frac{1}{2}$, $T > 1$. Let X be a large integer and

$$M(s, \chi) = \sum_{m \leqslant X} \mu(m)\chi(m)m^{-s}. \tag{23.2}$$

If Riemann's hypothesis were true, then for $\sigma > \frac{1}{2}$ $M(s, \chi)$ would tend to $\{L(s, \chi)\}^{-1}$ as X tended to infinity. We write

$$L(s, \chi)M(s, \chi) = 1 + f(s, \chi), \tag{23.3}$$

where $f(s, \chi)$ has the Dirichlet series

$$\sum_{m > X} a(m)\chi(m)m^{-s}, \tag{23.4}$$

$$a(m) = \sum_{\substack{d \mid m \\ d \leqslant X}} \mu(d). \tag{23.5}$$

The series (23.4) converges without any hypothesis for $\sigma > 1$. Until A. I. Vinogradov's work (1965), the number of zeros was customarily estimated by integration of the logarithm of $1 + f(s, \chi)$ round the boundary of the rectangle (23.1). As we shall see below, $f(s, \chi)$ is less than unity in root mean square, whether averaged over t or over χ or over both, provided $\sigma > \frac{1}{2}$. At a zero, $f(\rho, \chi)$ is -1. A. I. Vinogradov revived an alternative approach: to count directly the number of times $f(s, \chi)$ has modulus at least unity. While no simpler in detail, this method is the more flexible, and we follow it here.

The integral transform

$$\frac{1}{2\pi i} \int_{2-i\infty}^{2+i\infty} \Gamma(w)\left(\frac{Y}{m}\right)^{w} dw = e^{-m/Y} \tag{23.6}$$

is easily verified by moving the line of integration to $\operatorname{Re} w = \frac{1}{2}-R$, where R is a large positive integer, and letting R tend to infinity: the residues at poles of $\Gamma(w)$ give the terms of the exponential series. Hence

$$e^{-1/Y} + \sum_{m>X} a(m)\chi(m)m^{-\rho}e^{-m/Y}$$
$$= \frac{1}{2\pi i} \int_{2-i\infty}^{2+i\infty} L(\rho+w, \chi)M(\rho+w, \chi)Y^{w}\Gamma(w)\, dw. \tag{23.7}$$

Here ρ is a zero of $L(s, \chi)$ so that the zero of $L(\rho+w, \chi)$ cancels the pole of $\Gamma(w)$ at $w = 0$. We take the integral back to the line $\operatorname{Re} w = \frac{1}{2}-\beta$. If χ is not trivial, the right-hand side is equal to

$$\frac{1}{2\pi i} \int_{\frac{1}{2}-\beta-i\infty}^{\frac{1}{2}-\beta+i\infty} L(\rho+w, \chi)M(\rho+w, \chi)Y^{w}\Gamma(w)\, dw, \tag{23.8}$$

and if $\chi \bmod q$ is trivial there is an extra term

$$\varphi(q)q^{-1}M(1, \chi)Y^{1-\rho}\Gamma(1-\rho) \tag{23.9}$$

from the pole at $\rho+w = 1$.

We write
$$l = \log QT \tag{23.10}$$
and suppose that
$$\log X \leqslant 10l, \qquad \log Y \leqslant 10l. \tag{23.11}$$

The terms on the left of (23.7) with $m > 100lY$ contribute less than $\frac{1}{10}$, if l is sufficiently large. We recall Stirling's formula in the form (20.3), valid if $\lambda \ll |\tau|^{\frac{1}{2}}$:

$$|\Gamma(\lambda+i\tau)| = e^{-\frac{1}{2}\pi|\tau|}|\lambda+i\tau|^{\lambda-\frac{1}{2}}\{(2\pi)^{\frac{1}{2}}+O(|\tau|^{-\frac{1}{2}})\}. \tag{23.12}$$

The term (23.9) is now seen to be at most $\frac{1}{10}$ when $|\gamma| \geqslant 100l$, again provided l is sufficiently large. Apart from the zeros of $\zeta(s)$ with $|\gamma| \leqslant 100l$, all zeros fall into one or both of the following classes.

Class (i). Zeros ρ with

$$\left|\sum_{X<m\leqslant 100lY} a(m)\chi(m)m^{-\rho}e^{-m/Y}\right| > \tfrac{1}{3}. \tag{23.13}$$

Class (ii). Zeros ρ with

$$\left|\int_{\frac{1}{2}-\beta-i\infty}^{\frac{1}{2}-\beta+i\infty} L(\rho+w,\chi)M(\rho+w,\chi)Y^w\Gamma(w)\,dw\right| > \tfrac{2}{3}\pi. \tag{23.14}$$

We subdivide class (i) by writing the range $(X, 100lY]$ as the union of intervals I_r: $2^rY < m \leqslant 2^{r+1}Y$ (the first and last intervals having instead the end points X and $100lY$). A zero is of class (i, r) if

$$\left|\sum_{m\in I_r} a(m)\chi(m)m^{-\rho}e^{-m/Y}\right| > \{20(r^2+1)\}^{-1}. \tag{23.15}$$

By (12.19) or (12.20) the number of zeros of a fixed function $L(s,\chi)$ in a subrectangle $t \leqslant \gamma \leqslant t+1$, $\alpha \leqslant \beta \leqslant 1$ of (23.1) is $O(l)$. From each class of zeros in (23.1) we can pick out a sequence of zeros whose imaginary parts differ by at least unity, in such a way that the sequence contains a proportion of at least $\gg l^{-1}$ of the zeros of $L(s,\chi)$ of that class in (23.1). It is possible that our sequence contains all ρ at which $L(\rho,\chi)$ has a zero of the given class, but these ρ are multiple zeros of order $O(l)$. We write $N_1(\chi)$ for the number of class (i) zeros of $L(s,\chi)$ in (23.1), $N_2(\chi)$ for class (ii) zeros, and $N(\chi)$ for the total number. We call our subsequence the *representative zeros.*

Summing over representative zeros ρ of class (i, r) we have

$$\begin{aligned}\sum_{q\leqslant Q}\frac{q}{\varphi(q)}\sideset{}{^*}\sum_{\chi \bmod q}\sum_{\rho}\left|\sum_{m\in I_r}\frac{a(m)\chi(m)}{m^\rho}e^{-m/Y}\right|^2 &\ll l\sum_{m\in I_r}(m+Q^2T)d^2(m)m^{-2\alpha}\exp(-2^r)\\ &\ll l(2^rY)^{1-2\alpha}(Q^2T+2^rY)\log^4 Y\exp(-2^r)\\ &\ll l^5\exp(-2^r)\left(Q^2T(2^rY)^{1-2\alpha}+(2^rY)^{2-2\alpha}\right),\end{aligned} \tag{23.16}$$

by (19.26) with $\delta = 1$; we have used (23.3) to estimate $a(m)$ by a divisor function, then (2.24) and partial summation.

Comparing (23.15) and (23.16) and adding l^2 for the zeros of $\zeta(s)$ with $|\gamma| \leqslant 100l$, we have

$$\begin{aligned}\sum_{q\leqslant Q}\frac{q}{\varphi(q)}\sideset{}{^*}\sum_{\chi \bmod q} N_1(\chi) &\ll l^2+l^6\sum_r (r^2+1)\exp(-2^r)\times\\ &\quad\times\{Q^2T(2^rY)^{1-2\alpha}+(2^rY)^{2-2\alpha}\} \ll Y^{2-2\alpha}l^6,\end{aligned} \tag{23.17}$$

where we have assumed $X \geqslant Q^2T$. We now choose

$$X = Q^2Tl. \tag{23.18}$$

There are several ways of treating zeros of class (ii). The ingenious work of Montgomery (1969*b*) makes much use of this flexibility. To obtain Ingham's theorem we raise the expression on the left of (23.14) to the four-thirds power and sum over representative zeros $\rho = \beta + i\gamma$ of class (ii). By Hölder's inequality we have

$$\sum_{q \leqslant Q} \frac{q}{\varphi(q)} \sideset{}{^*}\sum_{\chi \bmod q} \sum_{\rho} \left| \int\limits_{\frac{1}{2}-\beta-i\infty}^{\frac{1}{2}-\beta+i\infty} L(\rho+w,\chi) M(\rho+w,\chi) Y^{w} \Gamma(w)\, dw \right|^{\frac{4}{3}}$$

$$\ll Y^{2(1-2\alpha)/3} \left(\sum_{q \leqslant Q} \frac{q}{\varphi(q)} \sideset{}{^*}\sum_{\chi \bmod q} \sum_{\rho} \int\limits_{\frac{1}{2}-i\infty}^{\frac{1}{2}+i\infty} |L(u+i\gamma,\chi)|^{4} |\Gamma(u-\beta)|^{\frac{4}{3}}\, |du| \right)^{\frac{1}{3}} \times$$

$$\times \left(\sum_{q \leqslant Q} \frac{q}{\varphi(q)} \sideset{}{^*}\sum_{\chi \bmod q} \sum_{\rho} \int\limits_{\frac{1}{2}-i\infty}^{\frac{1}{2}+i\infty} |M(u+i\gamma,\chi)|^{2} |\Gamma(u-\beta)|^{\frac{4}{3}}\, |du| \right)^{\frac{2}{3}}. \qquad (23.19)$$

By (23.12) and (22.22) the first integral on the right of (23.19) is

$$\ll \int\limits_{0}^{1} Q^{2} T l^{5} \max_{\beta} (t+\beta-\tfrac{1}{2})^{-\frac{4}{3}}\, dt +$$

$$+ \int\limits_{1}^{\infty} Q^{2}(T+t)(l+\log t)^{5} \max_{\beta} (e^{-2\pi t/3} t^{-4\beta/3})\, dt$$

$$\ll (\alpha - \tfrac{1}{2})^{-\frac{1}{3}} Q^{2} T l^{5}. \qquad (23.20)$$

When we apply the hybrid sieve (19.26), the second integral on the right of (23.19) becomes

$$\ll l(\alpha-\tfrac{1}{2})^{-\frac{1}{3}} \sum_{m=1}^{X} \frac{(m+Q^{2}T)}{m} \frac{\log X}{\log(m+e)}$$

$$\ll l(\alpha-\tfrac{1}{2})^{-\frac{1}{3}} (X+Q^{2}Tl)$$

$$\ll (\alpha-\tfrac{1}{2})^{-\frac{1}{3}} Q^{2} T l^{2}, \qquad (23.21)$$

by (23.18). The right-hand side of (23.19) is

$$\ll (\alpha-\tfrac{1}{2})^{-\frac{1}{3}} Q^{2} T Y^{2(1-2\alpha)/3} l^{3}, \qquad (23.22)$$

and thus
$$\sum_{q \leqslant Q} \frac{q}{\varphi(q)} \sideset{}{^*}\sum_{\chi \bmod q} N_{2}(\chi) \ll (\alpha-\tfrac{1}{2})^{-\frac{1}{3}} Q^{2} T Y^{2(1-2\alpha)/3} l^{4}. \qquad (23.23)$$

We now choose
$$Y = (Q^{2} T l^{-2})^{3/(4-2\alpha)}, \qquad (23.24)$$

so that (23.17) and (23.23) together give

$$\sum_{q \leqslant Q} \frac{q}{\varphi(q)} \sideset{}{^*}\sum_{\chi \bmod q} N(\chi) \ll (\alpha-\tfrac{1}{2})^{-\frac{1}{3}} (Q^{2}T)^{3(1-\alpha)/(2-\alpha)} l^{6/(2-\alpha)} \qquad (23.25)$$

We have assumed that $100lY$ is greater than X; this is certainly true if

$$\alpha > \tfrac{1}{2}+\tfrac{9}{4}l^{-1}\log l. \tag{23.26}$$

If α is less than this bound we quote (12.19) or (12.20), which give an upper bound

$$N(\chi) \ll Tl, \tag{23.27}$$

so that

$$\sum_{q\leqslant Q}\frac{q}{\varphi(q)}\sum_{\chi \bmod q}{}^{*} N(\chi) \ll Q^2Tl \tag{23.28}$$

in any case. In fact, (23.27) and (23.28) give

$$\sum_{q\leqslant Q}\frac{q}{\varphi(q)}\sum_{\chi \bmod q}{}^{*} N(\chi)$$
$$\ll \min\{Q^2Tl, (\alpha-\tfrac{1}{2})^{-\frac{1}{3}}(Q^2T)^{3(1-\alpha)/(2-\alpha)}l^{6/(2-\alpha)}\}. \tag{23.29}$$

Ingham proved his theorem for the zeta function only, with no sum over χ; his result (1940) was

$$N \ll T^{3(1-\alpha)/(2-\alpha)}l^5, \tag{23.30}$$

where N refers to the zeros of the zeta function satisfying (23.1). (23.25) is stronger than (23.30) for α near $\frac{1}{2}$, but weaker for $\alpha > \frac{4}{5}$. We have lost a logarithm factor by taking representative zeros instead of counting according to multiplicity.

24

BOMBIERI'S THEOREM

'I see, I see,' said Pooh, nodding his head. 'Talking about large somethings,' he went on dreamily, 'I generally have a small something about now—about this time in the morning.'
I. 48

WE proved the prime-number theorem for arithmetic progressions $a \bmod q$ uniformly in $q \leqslant Q$, where

$$Q = (\log x)^N, \tag{24.1}$$

in the notation of Chapter 15. In 1965, Bombieri (1965) and A. I. Vinogradov (1965) proved independently that the prime-number theorem holds uniformly for almost all progressions with Q a little smaller than $x^{\frac{1}{2}}$. Bombieri's result was the following theorem.

THEOREM. *Given $A > 0$ we can find $B > 0$ such that for all large x*

$$\sum_{q \leqslant Q} \max_{a \bmod q}{}^* \max_{y \leqslant x} \left| \psi(y; q, a) - \frac{y}{\varphi(q)} \right| \leqslant \frac{x}{(\log x)^A}, \tag{24.2}$$

where
$$Q = x^{\frac{1}{2}}(\log x)^{-B} \tag{24.3}$$

and the asterisk indicates a restriction to reduced residue classes.

The right-hand side of (24.2) is smaller by a factor $(\log x)^{A+1}$ than the sum of the corresponding main terms. The proportion of progressions for which the error term in the prime-number theorem for the interval $1 \leqslant p \leqslant x$ is greater than $(\log x)^{-A}$ times the main term is $O\big((\log x)\big)^{-1}$. In calculations we often apply the prime-number theorem to several progressions, and its failure in a small number of cases affects only the error term in the eventual asymptotic formula. An example is the proof that every large even integer is the sum of a prime and an integer with at most three prime factors. It may be that (24.2) holds with some larger constant in place of $\frac{1}{2}$, but $\frac{1}{2}$ is the limit of present methods.

An upper bound for the left-hand side of (24.2) is

$$\sum_{q \leqslant Q} \frac{1}{\varphi(q)} \sum_{\substack{\chi \bmod q \\ \chi \text{ non-trivial}}} \max_{y \leqslant x} |\psi(y, \chi)| + \sum_{q \leqslant Q} \frac{1}{\varphi(q)} \max_{y \leqslant x} |\psi(y, \chi_0) - y|. \tag{24.4}$$

When we use eqn (17.15) to pass to proper characters, the sum in (24.4) does not exceed

$$\sum_{1\leqslant f\leqslant Q}\sum_{\chi \bmod f}\max_{y\leqslant x}|\psi(y,\chi)|\sum^{*}_{\substack{q\equiv 0(\bmod f)\\ q\leqslant Q}}\frac{1}{\varphi(q)}+$$

$$+\max_{y\leqslant x}|\psi(y)-y|\sum_{q\leqslant Q}\frac{1}{\varphi(q)}+\lambda^{2}Q, \tag{24.5}$$

where we have written $\lambda = \log x.$ (24.6)

The coefficient of the term in $\chi \bmod f$ in (24.5) is

$$\sum_{\substack{q\equiv 0(\bmod f)\\ q\leqslant Q}}\frac{1}{\varphi(q)}\leqslant\frac{1}{\varphi(f)}\sum_{m\leqslant x}\frac{1}{\varphi(m)}\ll\frac{\lambda}{\varphi(f)}. \tag{24.7}$$

Relations (16.22) and (24.3) assure us that the second and third terms on the right of (24.5) are

$$\ll x\lambda^{-A} \tag{24.8}$$

for x sufficiently large. Since $f\leqslant\lambda^{4B}$ is of the form (24.1), the Siegel–Walfisch inequality (17.14) implies that (24.8) is an upper bound for the terms in (24.5) with $f\leqslant\lambda^{4B}$; we are assuming x is sufficiently large.

For $q>\lambda^{4B}$ we combine (17.2) and (17.6) into

$$\psi(y,\chi) = -\sum\frac{y^{\rho}}{\rho}+O\left\{\left(\frac{y}{T}+y^{\frac{1}{4}}\right)\log^{2}q\,Ty\right\}, \tag{24.9}$$

where the sum is over zeros of $\xi(s,\chi)$ with $|\rho|>R$ and $|\gamma|<T$, R and T being chosen for each χ so that (17.3) and (17.4) hold; in particular the O-constant in (24.9) is independent of χ, T, or y and uniform in $\frac{1}{5}\leqslant R\leqslant\frac{1}{4}$. When we choose T to be a little smaller than x we have

$$|\psi(y,\chi)|\leqslant\sum_{\substack{|\rho|>\frac{1}{5}\\ |\gamma|<x}}\frac{x^{\beta}}{|\rho|}+O(x^{\frac{1}{4}}\log^{2}x) \tag{24.10}$$

whenever $y\leqslant x$, with an absolute O-constant. The error term in (24.10) contributes

$$O(Qx^{\frac{1}{4}}\log^{3}x) \tag{24.11}$$

to the sum (24.4).

We now divide the rectangle $0\leqslant\beta\leqslant 1$, $-T\leqslant\gamma\leqslant T$ into smaller rectangles, so that β is in one of the ranges $[0,\frac{1}{2}+(\log x)^{-1}]$,..., $[\frac{1}{2}+r(\log x)^{-1},\frac{1}{2}+(r+1)(\log x)^{-1}]$,..., and $|\gamma|$ in one of the ranges $[0,1]$, $[1,2]$,..., $[2^{r},2^{r+1}]$,.... Within each rectangle, $x^{\beta}/|\rho|$ varies by a bounded factor. Again, we divide the range $(\lambda^{4B},Q]$ for q into intervals

$P < q \leqslant 2P$. We use Ingham's theorem: the number of zeros $\rho = \beta + i\gamma$ counted with weight $q/\varphi(q)$ of functions $L(s, \chi)$ with χ proper mod q, $P < q \leqslant 2P$, $\alpha \leqslant \beta \leqslant \alpha + (\log x)^{-1}$, and $U \leqslant \gamma \leqslant 2U$ is

$$\ll \min\bigl(P^2 U\lambda, (P^2 U)^{3(1-\alpha)/(2-\alpha)}\lambda^6\bigr), \tag{24.12}$$

by (23.25) and (23.28). Each of these zeros contributes at most

$$\frac{ex^\alpha}{|\rho|}\frac{\lambda}{\varphi(q)} \ll \frac{\lambda x^\alpha}{PU}\frac{q}{\varphi(q)} \tag{24.13}$$

to the sum in (24.5). We multiply the expression in (24.12) by $\lambda x^\alpha/PU$ and sum over the appropriate sequence of values of PU. First, we have

$$\sum_U U^{3(1-\alpha)/(2-\alpha)-1} \ll \lambda \tag{24.14}$$

when $\alpha > \frac{1}{2}$ and the sum is over powers of two not exceeding x.

In the sum over P we distinguish two cases. The exponent of P in (24.12) is greater than unity for $\alpha < \frac{4}{5}$ and less than unity for $\alpha > \frac{4}{5}$. The values of P are the powers of two between $\frac{1}{2}\lambda^{4B}$ and Q; they are $O(\lambda)$ in number. Hence for $\alpha \geqslant \frac{4}{5}$ we have

$$\begin{aligned}\sum_P P^{6(1-\alpha)/(2-\alpha)-1} &\ll \lambda(\lambda^{4B})^{6(1-\alpha)/(2-\alpha)-1}\\ &\ll \lambda^{1+4B(6(1-\alpha)-1)}\\ &\ll x^{1-\alpha}\lambda^{1-4B},\end{aligned} \tag{24.15}$$

when we assume x to be sufficiently large. For $\frac{1}{2} < \alpha \leqslant \frac{4}{5}$ we have

$$\begin{aligned}\sum_P P^{6(1-\alpha)/(2-\alpha)-1} &\ll \lambda Q^{6(1-\alpha)/(2-\alpha)-1}\\ &\ll x^{3(1-\alpha)/(2-\alpha)-\frac{1}{2}}\lambda^{1-B}\\ &\ll x^{1-\alpha}\lambda^{1-B},\end{aligned} \tag{24.16}$$

where we have substituted $Q = x^{\frac{1}{2}}\lambda^{-B}$ from (24.3). The terms from zeros of L-functions formed with characters whose conductors exceed λ^{4B} are now seen to be

$$\ll x\lambda^{10-B} \tag{24.17}$$

by (24.12), (24.13), (24.14), (24.15), and (24.16). The other terms in (24.5) are of the form (24.17), by (24.8) and (24.11), when we choose

$$B = A + 10. \tag{24.18}$$

We have thus proved (24.2) with B given by (24.18), which clearly can be improved slightly, since (24.15) and (24.16) can be improved.

There are two ways of proving Bombieri's theorem. The shorter, due to Gallagher (1968), is to perform the sieving directly on $L'(s, \chi)/L(s, \chi)$

and related functions in contour integrals. The proof is an anagram of the one we used; multiplication by $M(s, \chi)$ and fourth-power averages of $L(s, \chi)$ occur in it. The longer way is to prove a zero-density theorem and deduce Bombieri's theorem from that, as in this chapter. In either case, the Siegel–Walfisch form of the prime-number theorem has to be used, and the constants are non-effective. Without using the hybrid sieve or the approximate functional equation, we should obtain a weaker zero-density result than (23.29), but one still powerful enough to enable (24.2) to be deduced.

25

I. M. VINOGRADOV'S ESTIMATE

If we are to capture Baby Roo, we must get a Long Start, because Kanga runs faster than any of Us, even Me.

I. 93

THE sum
$$S(\alpha) = S(y, \alpha) = \sum_{p \leqslant y} \log p \, e(p\alpha) \tag{25.1}$$
is of the type considered in Chapter 6, with local maxima near rational points a/q (if y is sufficiently large). In this chapter, we adapt the proof of Bombieri's theorem to show rigorously that $S(\alpha)$ is small except possibly when α is close to a rational point with small denominator. More precisely, if B is fixed and x is large, then
$$|S(x, a/q+\beta)| \ll (1+x|\beta|)xl^{8-\frac{1}{2}B}, \tag{25.2}$$
where
$$l = \log x \tag{25.3}$$
and a/q is in its lowest terms with
$$l^B \leqslant q \leqslant xl^{-B}. \tag{25.4}$$
Inequality (25.2) is I. M. Vinogradov's famous 'minor arcs estimate', with 8 in place of Vinogradov's $\frac{7}{2}$. His proof is elementary, making no use of Dirichlet's series; an analytic proof was published by Linnik (1945).

We remark first that
$$S(y, \alpha) = \sum_{m \leqslant y} \Lambda(m)e(m\alpha) + O(y^{\frac{1}{2}} \log y), \tag{25.5}$$
the error term arising from squares and higher powers of primes. When a/q is in its lowest terms,
$$\sum_{m \leqslant y} \Lambda(m) e\left(\frac{ma}{q}\right) = \sideset{}{^*}\sum_{b \bmod q} e_q(ab) \sum_{\chi \bmod q} \frac{\bar{\chi}(b)}{\varphi(q)} \psi(y, \chi) + O(\log y \log q), \tag{25.6}$$
the error term arising from powers of the $O(\log q)$ primes that divide q; these are not congruent to reduced residues $b \bmod q$. The sum of the terms involving b in (25.6) is $\bar{\chi}(a)\tau(\bar{\chi})$, and by eqn (3.20) the modulus

of this expression is at most $q^{\frac{1}{2}}$. We combine (24.10) and (17.15) into

$$|\psi(y,\chi)| \leqslant \sum_{\substack{|\rho| \geqslant \frac{1}{5} \\ |\gamma| < x}} \frac{x^{\beta}}{|\rho|} + O(x^{\frac{1}{2}}l^2) \tag{25.7}$$

whenever $y \leqslant x$, the O-constant being absolute. The sum is over zeros of $\xi(s,\chi_1)$, where χ_1 proper mod f induces χ mod q. Since every zero of $\xi(s,\chi_1)$ is a zero of $L(s,\chi)$, we shall take the sum in (25.7) to be over zeros of $L(s,\chi)$. From (25.5), (25.6), and (25.7), we have

$$|S(y,a/q)| \leqslant \frac{q^{\frac{1}{2}}}{\varphi(q)} \sum_{\chi \bmod q} \sum_{\substack{|\rho| \geqslant \frac{1}{5} \\ |y| < x}} \frac{x^{\beta}}{|\rho|} + O(q^{\frac{1}{2}}x^{\frac{1}{2}}l^2), \tag{25.8}$$

where $y \leqslant x$ and $q \leqslant x$.

We do not use Bombieri's theorem itself, but analogous results for zeros of the functions $L(s,\chi)$ where χ induces a character to the fixed modulus q. This argument is not as natural as that leading to Bombieri's theorem, and the various inequalities do not fit together so well. First we give a form of Ingham's theorem for the set of all characters χ mod q, where q is fixed. If χ mod q is induced by χ_1 proper mod f, we have

$$L(s,\chi) = L(s,\chi_1) \prod_{\substack{p|q \\ p\nmid f}} \left(1 - \frac{\chi_1(p)}{p^s}\right). \tag{25.9}$$

The method of Chapter 22 with (19.24) instead of (19.26) gives

$$\sum_{\chi \bmod f}^{*} \sum_{r=1}^{R(\chi)} |L(\tfrac{1}{2}+it(r,\chi))|^4 \leqslant \varphi(f)T\log^5 fT, \tag{25.10}$$

and hence

$$\begin{aligned}\sum_{\chi \bmod q} \sum_{r=1}^{R(\chi)} |L(\tfrac{1}{2}+it(r,\chi))|^4 &\leqslant T\log^5 qT \sum_{f|q} \varphi(f) \prod_{\substack{p|q \\ p\nmid f}} \left(1+\frac{1}{p^{\frac{1}{2}}}\right) \\ &\leqslant T\log^5 qT \prod_{p|q} (p-1+1+p^{-\frac{1}{2}}) \\ &\leqslant qT\log^5 qT, \end{aligned} \tag{25.11}$$

since the product over primes dividing q is at most $q\zeta(\frac{3}{2})$. Here we suppose that the points $t(r,\chi)$ satisfy conditions (22.23) and (22.24).

When we use (18.8) and (25.11) in place of (18.12) and (22.22), Ingham's theorem becomes: the number of zeros $\rho = \beta+i\gamma$ of functions $L(s,\chi)$ formed with characters χ mod q in the rectangle $\alpha \leqslant \beta \leqslant 1$, $-T \leqslant \gamma \leqslant T$ is at most

$$\leqslant \max\{\varphi(q)T\lambda, (\alpha-\tfrac{1}{2})^{-\frac{1}{3}}(qT)^{3(1-\alpha)/(2-\alpha)}(\varphi(q)\lambda^6/q)^{1/(2-\alpha)}\} \tag{25.12}$$

where
$$\lambda = \log qT. \tag{25.13}$$

As in the proof of Bombieri's theorem, we sum zeros over rectangles. For $\alpha \geqslant \frac{4}{5}$ the analogue of (24.15) is

$$q^{3(1-\alpha)/(2-\alpha)-\frac{1}{2}} \ll (l^B)^{3(1-\alpha)/(2-\alpha)-\frac{1}{2}} \ll x^{1-\alpha}l^{-\frac{1}{2}B} \tag{25.14}$$

when x is sufficiently large, and for $\alpha \leqslant \frac{4}{5}$ the analogue of (24.16) is

$$q^{3(1-\alpha)/(2-\alpha)-\frac{1}{2}} \ll (xl^{-B})^{3(1-\alpha)/(2-\alpha)-\frac{1}{2}} \ll x^{1-\alpha}l^{-\frac{1}{2}B}. \tag{25.15}$$

Since the primes less than q and not dividing it are reduced mod q, we see from the prime-number theorem that $\varphi(q)\lambda^6/q$ is greater than unity for q satisfying (25.4). By (25.14), (25.15), and (24.14) the sum over χ and ρ in (25.8) is

$$\ll q^{\frac{1}{2}}xl^{2-\frac{1}{2}B}\varphi(q)l^6/q, \tag{25.16}$$

and we have proved (25.2) for $\beta = 0$.

We now show that (25.2) holds (with a different O-constant) when β is different from zero. Since

$$\int_0^m 2\pi i\beta\, e(\beta y)\, dy = e(m\beta)-1, \tag{25.17}$$

we have the identity

$$S(x, a/q+\beta)$$
$$= S(x, a/q)+2\pi i\beta \int_0^x \{S(x, a/q)-S(y, a/q)\}e(\beta y)\, dy. \tag{25.18}$$

The right-hand side of eqn (25.18) is in modulus

$$\leqslant (1+4\pi|\beta|x)\max_{y\leqslant x}|S(y, a/q)|, \tag{25.19}$$

and we have proved (25.2).

26

I. M. VINOGRADOV'S THREE-PRIMES THEOREM

WITH $S(y, \alpha)$ given by eqn (25.1) and x an integer we have

$$\int_0^1 S^3(x,\alpha)e(-x\alpha)\,d\alpha = \sum_{p_1}\sum_{p_2}\sum_{p_3} \log p_1 \log p_2 \log p_3, \tag{26.1}$$

where the sum is over triples (p_1, p_2, p_3) of primes with

$$p_1+p_2+p_3 = x. \tag{26.2}$$

We shall find an asymptotic formula for the left-hand side of eqn (26.1), and deduce that every sufficiently large odd integer is a sum of three primes. Since we have to use the prime-number theorem for arithmetical progressions, the constant in the error term of the asymptotic formula depends on Siegel's theorem and cannot be stated; however, there are weaker results than Siegel's theorem in which the constants are effective, and in principle a finite set could be given in which any exceptional x must lie.

Let

$$Q = [xl^{-28}], \tag{26.3}$$

where $l = \log x$. Since $S(x,\alpha)$ is periodic, we can take the range of integration in (26.1) to be the unit interval $[1/(Q+1), (Q+2)/(Q+1)]$. We divide this interval into arcs

$$I_r = \left[\frac{a_r+a_{r-1}}{q_r+q_{r-1}}, \frac{a_r+a_{r+1}}{q_r+q_{r+1}}\right] \tag{26.4}$$

corresponding to the fractions a_r/q_r of the Farey sequence of order Q. By eqn (9.2) if $a_r/q_r+\beta$ is on I_r then

$$|\beta| \leqslant (q_r Q)^{-1}. \tag{26.5}$$

The intervals I_r are known as *Farey arcs* (if we think of $e(\alpha)$ as a complex variable, the integration is round the unit circle). We divide the I_r into *major arcs,* on which we work out the integral over the spike at a_r/q_r explicitly, as promised in Chapter 6, and *minor arcs,* on which we use an upper estimate for the integrand. We call I_r a minor arc if $q_r > l^{26}$, when (25.2) with $B = 26$ gives

$$|S(x, a_r/q_r+\beta)| \ll (1+xl^{-26}Q^{-1})xl^{-6} \ll xl^{-4} \tag{26.6}$$

for a_r/q_r lying on I_r.

On the major arcs we approximate $S(\alpha)$. First we calculate the size of the spike at a/q. From (17.19) we have

$$\left|\sum_{\substack{p\leqslant y\\ p\equiv b(\mathrm{mod}\, q)}} \log p - \frac{[y]}{\varphi(q)}\right| \ll xl^{-32} \tag{26.7}$$

uniformly in $q \leqslant l^{26}$ and in b reduced mod q. We recall that

$$\sum_{b \bmod q}^{*} e_q(ab) = c_q(a) = \mu(q) \tag{26.8}$$

when a is reduced mod q. Hence

$$|S(y, a/q)-\mu(q)[y]/\varphi(q)| \leqslant l+\sum_{b \bmod q}^{*}\left|\sum_{\substack{p\leqslant y\\ p\equiv b(\mathrm{mod}\, q)}} \log p - \frac{[y]}{\varphi(q)}\right|$$
$$\ll \varphi(q)xl^{-32}. \tag{26.9}$$

The simplest sum with a spike is

$$F(\alpha) = \sum_{m\leqslant x} e(m\alpha). \tag{26.10}$$

By the identity (25.18),

$$\left|S\left(x, \frac{a}{q}+\beta\right)-\frac{\mu(q)}{\varphi(q)}F(\beta)\right| \ll (1+x|\beta|)\max_{y\leqslant x}\left|S\left(y,\frac{a}{q}\right)-\frac{\mu(q)}{\varphi(q)}[y]\right|$$
$$\ll l^{28}q^{-1}\varphi(q)xl^{-32} \ll xl^{-4}. \tag{26.11}$$

We now have

$$\left|S^3(a/q+\beta)-\frac{\mu^3(q)}{\varphi^3(q)}F^3(\beta)\right| \ll xl^{-4}\left(|S(a/q+\beta)|^2+\frac{\mu^2(q)}{\varphi^2(q)}|F(\beta)|^2\right), \tag{26.12}$$

and the integral in eqn (26.1) is

$$\sum_{I_r \text{ major}} \int_{I_r} \frac{\mu^3(q_r)}{\varphi^3(q_r)} F^3(\alpha - a_r/q_r)e(-x\alpha)\,\mathrm{d}\alpha + O\left(xl^{-4}\int_{1/(Q+1)}^{(Q+2)/(Q+1)} |S(\alpha)|^2\,\mathrm{d}\alpha\right)+$$
$$+O\left(\sum_{I_r \text{ major}} xl^{-4}\frac{\mu^2(q)}{\varphi^2(q)}\int_{I_r}\left|F\left(\alpha-\frac{a_r}{q_r}\right)\right|^2\mathrm{d}\alpha\right). \tag{26.13}$$

The integral in the first error term in (26.13) is

$$\sum_{p\leqslant x} \log^2 p \ll xl \tag{26.14}$$

by the prime-number theorem or by (8.19), and the integral in the second error term is

$$\ll \int_{-\frac{1}{2}}^{\frac{1}{2}} |F(\beta)|^2 \, d\beta = x. \tag{26.15}$$

Using (8.16) for $\sum \mu^2(q)/\varphi(q)$, we see that the error terms in (26.13) are both

$$\ll xl^{-3}. \tag{26.16}$$

Finally we must work out the first integral in (26.13). Since

$$F(\alpha) = \frac{e(x\alpha+\alpha)-1}{e(\alpha)-1} \ll \frac{1}{\|\alpha\|}, \tag{26.17}$$

we have

$$\left| e_{q_r}(-a_r x) \int_{-\frac{1}{2}}^{\frac{1}{2}} F^3(\beta)e(-x\beta)\, d\beta - \int_{I_r} F^3\left(\alpha - \frac{a_r}{q_r}\right) e(-x\alpha)\, d\alpha \right|$$

$$\ll \int_{l^2x^{-1}}^{\frac{1}{2}} \frac{d\beta}{\beta^3} \ll x^2 l^{-4} \tag{26.18}$$

when $q_r \leqslant l^{26}$. The integral over β on the left of (26.18) is

$$\tfrac{1}{2}x(x-1), \tag{26.19}$$

and the integral in eqn (26.1) is now seen to be

$$\tfrac{1}{2}x(x-1) \sum_{q \leqslant l^{26}} \sideset{}{^*}\sum_{a \bmod q} \frac{e_q(-ax)\mu^3(q)}{\varphi^3(q)} + O\left(\frac{x^2}{l^3}\right). \tag{26.20}$$

We can replace the sum over $q \leqslant l^{26}$ by one over q from 1 to infinity with an error term $O(x^2 l^{-26})$. Now

$$\sum_{q=1}^{\infty} \sideset{}{^*}\sum_{a \bmod q} \frac{e_q(-ax)\mu^3(q)}{\varphi^3(q)} = \sum_q \frac{\mu^3(q)c_q(-x)}{\varphi^3(q)}$$

$$= \sum_q \sum_{\substack{d|x \\ d|q}} \frac{d\mu(q/d)\mu^3(q)}{\varphi^3(q)}$$

$$= \sum_{d|x} d\mu(d) \sum_{q \equiv 0 (\bmod d)} \frac{\mu^2(q)}{\varphi^3(q)}$$

$$= \prod_p \left(1 + \frac{1}{(p-1)^3}\right) \prod_{p|x} \left(1 - \frac{p}{(p-1)^3} \frac{(p-1)^3}{(p-1)^3+1}\right)$$

$$= \prod_p \left(1 + \frac{1}{(p-1)^3}\right) \prod_{p|x} \left(1 - \frac{1}{p^2-3p+3}\right), \tag{26.21}$$

which is zero if x is even and positive if x is odd. We put $A(x)$ for the constant in (26.21), so that the left-hand side of (26.1) is

$$\tfrac{1}{2}A(x)x^2+O(x^2l^{-3}), \tag{26.22}$$

which is non-zero for x odd and sufficiently large.

Since primes less than xl^{-2} can contribute at most x^2l^{-1} to the sum on the right of eqn (26.1), and

$$l \geqslant \log p \geqslant l-2\log l \tag{26.23}$$

if $x \geqslant p \geqslant xl^{-2}$, we can assert that the number of solutions of (26.2) is

$$\tfrac{1}{2}A(x)x^2l^{-3}+O(x^2l^{-4}\log l), \tag{26.24}$$

an expression which again is seen to be non-zero for x odd and sufficiently large. The formula (26.24) is the famous theorem of I. M. Vinogradov.

27

HALÁSZ'S METHOD

'And I know it *seems* easy,' said Piglet to himself, 'but it isn't *everyone* who could do it.'

II. 18

WE have been concerned with estimating how often various sums S are large. The large-sieve results that we have had gave an upper bound for the sum of $|S|^2$ over different values of a parameter. The upper bounds we obtained contained two terms; the first was the maximum of $|S|^2$ and the second the mean square of $|S|$ multiplied by the number of values of the parameter. Thus if

$$S(t) = \sum_{m=1}^{N} a(m)m^{it} \tag{27.1}$$

then (18.29) gives

$$\sum_{r=1}^{R} |S(t_r)|^2 \ll N\log^2 N \sum |a(m)|^2 + (T_2-T_1)\log N \sum |a(m)|^2, \tag{27.2}$$

where the points t_r are at least $(\log N)^{-1}$ apart and lie between T_1 and T_2. The second term in (27.2) corresponds to the mean value of $|S(t)|^2$ and the first to its maximum. Halász's method sometimes enables us to count the number of values of r for which $|S(t_r)|$ is large, without the second term's being present on the right of (27.2).

Halász's method is based on the lemma of Chapter 7,

$$\Big|\sum_{r=1}^{R} c_r(\mathbf{u},\mathbf{f}^{(r)})\Big| \leqslant \|\mathbf{u}\|\Big(\sum_{1}^{R}|c_r|^2\Big)^{\frac{1}{2}}\Big(\max_{1\leqslant r\leqslant R}\sum_{q=1}^{R}|(\mathbf{f}^{(r)},\mathbf{f}^{(q)})|^2\Big)^{\frac{1}{4}}. \tag{27.3}$$

In the proof of (27.3), we assumed that the vectors $\mathbf{u}$, $\mathbf{f}^{(1)},\ldots,\mathbf{f}^{(R)}$ had finite dimension N, but the proof remains valid when the dimension is infinite, provided that each vector involved has finite norm. We choose the coefficients c_r so that $c_r(\mathbf{u},\mathbf{f}^{(r)})$ is real and positive. For (7.15) we took c_r to be the complex conjugate of $(\mathbf{u},\mathbf{f}^{(r)})$, but here we give c_r unit modulus, so that the second factor on the right of (27.3) is $R^{\frac{1}{2}}$. If each term on the left of (27.3) is at least V we can square (27.3) to give

$$R^2V^2 \leqslant R\|\mathbf{u}\|^2 \max_{1\leqslant r\leqslant R} \sum_{q=1}^{R} |(\mathbf{f}^{(r)}, \mathbf{f}^{(q)})|$$

$$\leqslant R\|\mathbf{u}\|^2\Big(\max_r \|\mathbf{f}^{(r)}\|^2 + (R-1)\max_{\substack{r,q\\ r\neq q}} |(\mathbf{f}^{(r)}, \mathbf{f}^{(q)})|\Big). \tag{27.4}$$

We can now deduce Halász's lemma.

LEMMA. *Let* $\mathbf{u}, \mathbf{f}^{(1)},\ldots, \mathbf{f}^{(R)}$ *be vectors of finite norm with*

$$|(\mathbf{u}, \mathbf{f}^{(r)})| \geqslant V \tag{27.5}$$

for $r = 1, 2,\ldots, R$. *Then*

$$R \leqslant 2V^{-2}\|\mathbf{u}\|^2 \max_{1\leqslant r\leqslant R} \|\mathbf{f}^{(r)}\|^2 \tag{27.6}$$

provided that

$$V^2 \geqslant 2\|\mathbf{u}\|^2 \max_{q\neq r} |(\mathbf{f}^{(r)}, \mathbf{f}^{(q)})|. \tag{27.7}$$

We shall apply (27.6) with

$$(\mathbf{u}, \mathbf{f}^{(r)}) = \sum_{m\leqslant N} a(m)m^{-s_r}, \tag{27.8}$$

where $s_r = \sigma_r + \mathrm{i}t_r$, $0 \leqslant \sigma_r \leqslant \frac{1}{3}$. Following Montgomery (1969*a*), we take

$$\left.\begin{aligned} u_m &= \mathrm{e}^{m/N}a(m) && \text{if } 1 \leqslant m \leqslant N \\ u_m &= 0 && \text{if } m > N \end{aligned}\right\}, \tag{27.9}$$

$$f_m^{(r)} = \mathrm{e}^{-m/N}m^{-\sigma_r - \mathrm{i}t_r} \quad \text{for all } m \geqslant 1. \tag{27.10}$$

Here,

$$\|\mathbf{u}\|^2 \leqslant \mathrm{e}^2 \sum_{1}^{N} |a(m)|^2 \tag{27.11}$$

and

$$\|\mathbf{f}^{(r)}\|^2 \leqslant \sum_{1}^{\infty} \mathrm{e}^{-2m/N} = \tfrac{1}{2}N + O(1). \tag{27.12}$$

Writing σ for $\sigma_r + \sigma_q$ and t for $t_q - t_r$, we have

$$(\mathbf{f}^{(r)}, \mathbf{f}^{(q)}) = \sum_{m=1}^{\infty} \mathrm{e}^{-2m/N}m^{-\sigma-\mathrm{i}t}$$

$$= \frac{1}{2\pi\mathrm{i}} \int_{2-\mathrm{i}\infty}^{2+\mathrm{i}\infty} \Gamma(w)(\tfrac{1}{2}N)^w \zeta(w+\sigma+\mathrm{i}t)\, \mathrm{d}w \tag{27.13}$$

by the integral transform (23.6).

Before estimating the integral in eqn (27.13), we move the line of integration to the contour C consisting of

C_1: the line segment $(-\mathrm{i}\infty, -\mathrm{i}(\log N)^{-1}]$,
C_2: the semicircle, centre the origin, radius $(\log N)^{-1}$, to the right of the imaginary axis,
C_3: the line segment $[\mathrm{i}(\log N)^{-1}, \mathrm{i}\infty)$.

A residue
$$\Gamma(1-\sigma-\mathrm{i}t)(\tfrac{1}{2}N)^{1-\sigma-\mathrm{i}t} \tag{27.14}$$
accrues from the pole of $\zeta(w+\sigma+\mathrm{i}t)$ at $w+\sigma+\mathrm{i}t = 1$. We recall Stirling's formula in the form (20.3):
$$|\Gamma(\lambda+\mathrm{i}\tau)| = \mathrm{e}^{-\frac{1}{2}\pi|\tau|}|\lambda+\mathrm{i}\tau|^{\lambda-\frac{1}{2}}\{(2\pi)^{\frac{1}{2}}+O(|\tau|^{-\frac{1}{2}})\}, \tag{27.15}$$
valid when $\lambda \ll |\tau|^{\frac{1}{2}}$. Hence if
$$|t| > \log N \tag{27.16}$$
the residue (27.14) is bounded.

Next we need a bound for $|\zeta(\lambda+\mathrm{i}\tau)|$. From the approximate functional equation of Chapter 21 we have for $0 \leqslant \lambda \leqslant \frac{1}{2}$, $|\tau| \geqslant 10$
$$\begin{aligned}|\zeta(1-\lambda-\mathrm{i}\tau)|^2 &\ll \sum_{m\leqslant O(|\tau|)} d(m)m^{\lambda-1}+ \\ &\quad + \max_{|r|\leqslant 2}\left|\frac{\Gamma\{\frac{1}{2}(\lambda+\mathrm{i}\tau+r\pi\mathrm{i})\}}{\Gamma\{\frac{1}{2}(1-\lambda-\mathrm{i}\tau-r\pi\mathrm{i})\}}\right| \sum_{m\leqslant O(|\tau|)} \frac{d(m)}{m^{\lambda}}+|\tau|^{-5} \\ &\ll |\tau|^{\lambda}\log^2|\tau|,\end{aligned} \tag{27.17}$$
where we have used (20.4) and (2.14), and partial summation. The functional equation and the estimate (20.4) now give
$$|\zeta(\lambda+\mathrm{i}\tau)|^2 \ll |\tau|^{1-\lambda}\log^2|\tau|, \tag{27.18}$$
and (27.17) and (27.18) together give
$$|\zeta(\lambda+\mathrm{i}\tau)| \ll |\tau|^{\frac{1}{2}(1-\lambda)}\log|\tau| \tag{27.19}$$
uniformly in $0 \leqslant \lambda \leqslant 1$, $|\tau| \geqslant 10$.

We have now shown that
$$\begin{aligned}&|(\mathbf{f}^{(r)}, \mathbf{f}^{(q)})| \\ &\ll 1+\int_C |\Gamma(w)||\zeta(w+\lambda+\mathrm{i}\tau)||\tfrac{1}{2}N|^{w}\,|\mathrm{d}w| \\ &\ll 1+|t|^{\frac{1}{2}}\log|t|\log N+\int_2^{\infty} \mathrm{e}^{-\frac{1}{2}\pi\tau}\tau^{-\frac{1}{2}}(1+|\tau-t|^{\frac{1}{2}}\log(|\tau-t|+e))\,\mathrm{d}\tau \\ &\ll |t|^{\frac{1}{2}}\log^2 N|t|,\end{aligned} \tag{27.20}$$

where we have used (27.15) for the gamma function and (27.19) for the zeta function. We can now restate the condition (27.8) as

$$T \leqslant T_0, \tag{27.21}$$

where T_0 is such that equality holds in (27.8) when we use (27.20) in substituting for the scalar product. Hence

$$V^2 \gg \sum_1^N |a(m)|^2 T_0^{\frac{1}{3}} \log^2 NT. \tag{27.22}$$

Clearly when $T > T_0$ we must divide up the range for T into intervals of length at most T_0. Repeated application of the inequality (27.7) gives us

$$R \ll \left(\frac{T}{T_0}+1\right) V^{-2} \sum_{m=1}^N |a(m)|^2 N. \tag{27.23}$$

When we substitute for T_0 we have the result which follows:

THEOREM. *Let*

$$G = \sum_{m=1}^N |a(m)|^2. \tag{27.24}$$

If

$$\left|\sum_{m=1}^N a(m) m^{-s}\right| \geqslant V \tag{27.25}$$

for $s = s_1, \ldots, s_R$, *where* $s_r = \sigma_r + it_r$ *with* $0 \leqslant \sigma_r \leqslant \frac{1}{3}$ *and*

$$T \geqslant |t_r - t_q| \geqslant \log N \tag{27.26}$$

for $q \neq r$, *then*

$$R \ll GNV^{-2} + G^3 NTV^{-6} \log^4 NT, \tag{27.27}$$

the implied constants being absolute.

The form of the second term in (27.27) arises from our choice of functions $\mathbf{f}^{(r)}$; it is larger than the first term unless (27.22) holds with T in place of T_0. A plausible conjecture is that

$$R \ll GNV^{-2} \tag{27.28}$$

whenever

$$V^2 \gg GT^\delta \tag{27.29}$$

for any fixed $\delta > 0$.

The use of the zeta function to prove (27.27) is a curious feature of Halász's method. If Lindelöf's hypothesis is true, we can take the line of integration in (27.13) to $\operatorname{Re} w = \frac{1}{2}$, with the effect of replacing $T_0^{\frac{1}{3}}$ in (27.22) by $N^{\frac{1}{2}} T_0^{\epsilon}$ for any $\epsilon > 0$. This is an improvement for $T > N$ (and if $T \leqslant N$ then (27.28) follows trivially from (27.2)), but it is still a long way from weakening the condition on T_0 to (27.29).

28

GAPS BETWEEN PRIME NUMBERS

'I shall do it', said Pooh, after waiting a little longer, 'by means of a trap. And it must be a Cunning Trap, so you will have to help me, Piglet.'

I. 56

FIRST we prove a theorem on the zeros of $\zeta(s)$, replacing the large sieve (19.26) by Halász's method in the work of Chapter 23. We shall use the notation of that chapter with $Q = 1$, so that only zeros of the zeta function are considered. The definition of class (i) and class (ii) zeros remains as before. We pick representatives of each class of zeros in such a way that their imaginary parts differ by at least $2l$, where

$$l = \log T, \tag{28.1}$$

but the representatives are in number $\gg l^{-2}$ times the zeros in that class. We suppose $\alpha \geqslant \frac{3}{4}$, since the result (28.19) which we obtain below improves on Ingham's theorem only for $\alpha > \frac{3}{4}$. The parameters X and Y will satisfy

$$X \leqslant T^2, \qquad 100lY \leqslant T^2. \tag{28.2}$$

In the definition (23.14) of a class (ii) zero $\rho = \beta + i\gamma$,

$$\left| \int_{\frac{1}{2}-\beta-i\infty}^{\frac{1}{2}-\beta+i\infty} \zeta(\rho+w)M(\rho+w)Y^w\Gamma(w)\,dw \right| > \tfrac{2}{3}\pi, \tag{28.3}$$

the parts of the integrand with $|\operatorname{Im} w| > 100l$ give less than $\frac{1}{2}$ (if l is sufficiently large). The integral of $|\Gamma(\frac{1}{2}+it)|$ converges rapidly so, for (28.3) to hold, there must be some t with $|t-\gamma| \leqslant 100l$ for which

$$|\zeta(\tfrac{1}{2}+it)M(\tfrac{1}{2}+it)| > cY^{\beta-\frac{1}{2}}, \tag{28.4}$$

where c is an absolute constant. We pick as representatives of the class (ii) zeros a sequence of values of t satisfying (28.4).

By (22.22) the number of these t with

$$|\zeta(\tfrac{1}{2}+it)| > U, \tag{28.5}$$

where we choose U below, is

$$\ll TU^{-4}l^5. \tag{28.6}$$

Otherwise we have

$$|M(\tfrac{1}{2}+it)| \geqslant V = cU^{-1}Y^{\alpha-\frac{1}{2}}, \tag{28.7}$$

and by (27.28) the number of such t is

$$\ll XV^{-2}l+XTV^{-6}l^{7}. \tag{28.8}$$

We choose

$$U = X^{-1/10}Y^{3(2\alpha-1)/10}, \tag{28.9}$$

$$V = cX^{1/10}Y^{(2\alpha-1)/5}, \tag{28.10}$$

and on adding (28.6) and (28.8) and multiplying by l^2 we see that class (ii) zeros number

$$\ll X^{2/5}Y^{-6(2\alpha-1)/5}Tl^{9}+X^{4/5}Y^{-2(2\alpha-1)/5}l^{3}, \tag{28.11}$$

the second term in (28.11) being less than the first provided

$$X^{2}Y^{4(2\alpha-1)} \ll T^{5}l^{30}. \tag{28.12}$$

A zero is of class (i, r) if

$$\Big|\sum_{m\in I_r} a(m)m^{-\rho}e^{-m/Y}\Big| > \{20(r^2+1)\}^{-1}. \tag{28.13}$$

We pick representatives and apply (27.28) with

$$G = \sum_{m\in I_r} |a(m)|^{2}m^{-2\alpha}e^{-2m/Y} \ll (2^{r}Y)^{1-2\alpha}\exp(-2^{r})l^{3}. \tag{28.14}$$

The number of representatives is thus

$$\ll r^{4}(2^{r}Y)^{2-2\alpha}\exp(-2^{r+1})l^{3}+r^{12}(2^{r}Y)^{4-6\alpha}T\exp(-3\,.\,2^{r+1})l^{13}. \tag{28.15}$$

Summing over r and multiplying by l^2, we see that there are at most

$$\ll Y^{2-2\alpha}l^{6}+X^{4-6\alpha}Tl^{27} \tag{28.16}$$

class (i) zeros. Choosing

$$X = T^{\frac{1}{2}(2\alpha-1)/(\alpha^2+\alpha-1)}, \tag{28.17}$$

$$Y = T^{\frac{1}{2}(5\alpha-3)/(\alpha^2+\alpha-1)}, \tag{28.18}$$

we find (28.12) is satisfied for $0 \leqslant \alpha \leqslant 1$, and that the number $N(\alpha, T)$ of zeros $\rho = \beta+i\gamma$ of $\zeta(s)$ with $\beta \geqslant \alpha$ and $|\gamma| \leqslant T$ satisfies the relation

$$N(\alpha, T) \ll T^{\{(5\alpha-3)(1-\alpha)\}/(\alpha^2+\alpha-1)}l^{27} \tag{28.19}$$

for $\frac{3}{4} \leqslant \alpha \leqslant 1$. The result (28.19) is also true for $\frac{1}{2} \leqslant \alpha \leqslant \frac{3}{4}$ by Ingham's theorem.

We now sketch the proof of our theorem on gaps between prime numbers.

THEOREM. *Let c be a real number greater than $\frac{7}{12}$. Then whenever x is sufficiently large, there is a prime p with*

$$x < p \leqslant x+x^{c}. \tag{28.20}$$

Such a result was first proved by Hoheisel (1930) with c a little less than one. Ingham (1937) obtained the result with $c > \frac{5}{8}$ and indicated how to replace $\frac{5}{8}$ with a smaller number by improving an upper bound for $|\zeta(\frac{1}{2}+it)|$. Several authors achieved this by means of intricate arguments. Recently Montgomery (1969*b*) obtained the result for $c > \frac{3}{5}$ by the method given here, but with a less efficient use of the Halász lemma; the improvement to $\frac{7}{12}$ was seen by the author in preparing the present exposition. As we have seen, Montgomery's method rests on the Halász lemma, and thus on bounds for $|\zeta(1+it)|$. As with Ingham's result $c > \frac{5}{8}$, improvements at $\frac{1}{2}+it$ improve the constant, in that a good estimate for the mean of a higher power than $|\zeta(\frac{1}{2}+it)|^4$ would decrease the estimate for class (ii) zeros, both in (28.19) and in Ingham's theorem. However, even if we knew Lindelöf's hypothesis, we should only be able to deduce (28.20) for $c > \frac{1}{2}$. It has long been conjectured (Cramér 1936) that for large x there is always a prime p with

$$x < p \leqslant x+O(\log^2 x), \tag{28.21}$$

but there seems no chance of approaching this conjecture by present methods.

There are two essentials for a proof of (28.20) with $c < 1$: a zero-density theorem such as (28.19) and a result on zeros of $\zeta(s)$ with β close to 1. We shall assume that $\zeta(s)$ has no zeros $\rho = \beta+i\gamma$ with

$$\beta \geqslant 1-A\{\log(|\gamma|+e)\}^{-B}, \tag{28.22}$$

where $B < 1$. The inequality (28.22) is proved by Hadamard's double-height method just as (13.12) was, but the proof uses such inequalities as

$$|\zeta(1+it)| \ll \log^c(|t|+e) \tag{28.23}$$

with $c < 1$. No better way to prove bounds for $|\zeta(1+it)|$ is known than to replace $\exp(-it\log m)$ by $\exp\{-itP(m)\}$, where $P(m)$ is a polynomial arising from the first few terms in the expansion of the logarithmic series. Since m runs through integer values, the resulting sum depends only on the fractional part of $\frac{1}{2}t/\pi$. More direct arguments fail, because t is much larger than any other parameter involved. After this transformation we must use the intricate methods of I. M. Vinogradov; Weyl's simpler approach gives only

$$|\zeta(\tfrac{1}{2}+it)| \ll \log(|t|+e)/\log\log(|t|+e^2), \tag{28.24}$$

where any higher power of $\log\log t$ than the first would suffice for the application. The proof of (28.22) and (28.23) occupies one and a half chapters of Titchmarsh (1951).

While proving the prime number theorem in Chapter 16 we saw that

$$\psi(x) = x - \sum_{|\gamma| < T} \frac{x^{\rho}}{\rho} + O\left(\frac{x \log^2 x}{T}\right), \tag{28.25}$$

where T is chosen less than x with the property that each zero $\rho = \beta + i\gamma$ of $\zeta(s)$ has

$$|\gamma - T| \gg \log T, \tag{28.26}$$

the sum in (28.25) being over all zeros ρ of $\xi(s, \chi)$ with $|\gamma| < T$. Hence

$$\psi(x+h) - \psi(x) = h + \sum_{|\gamma| < T} \frac{x^{\rho} - (x+h)^{\rho}}{\rho} + O\left(\frac{x\lambda^2}{T}\right), \tag{28.27}$$

where we have written

$$\lambda = \log x. \tag{28.28}$$

By the mean-value theorem the sum over zeros in (28.27) is in modulus

$$\leqslant \sum_{|\gamma| < T} h(x + \theta h)^{\beta - 1} \ll h \sum_{|\gamma| < T} x^{\beta - 1} \tag{28.29}$$

for some θ in $0 < \theta < 1$. To estimate the sum in (28.29) we divide the interval $[0, 1]$ into ranges $[0, \frac{1}{2}], [\frac{1}{2}, \frac{1}{2} + \lambda^{-1}], \ldots, [\frac{1}{2} + r\lambda^{-1}, \frac{1}{2} + (r+1)\lambda^{-1}], \ldots$. The number of zeros $\rho = \beta + i\gamma$ with $\alpha \leqslant \beta \leqslant \alpha + \lambda^{-1}$ is

$$\ll T^{\frac{12}{5}(1-\alpha)} \lambda^{27}, \tag{28.30}$$

this being by (28.19) for $\alpha \geqslant \frac{3}{4}$ and by Ingham's theorem (23.29) for $\frac{1}{2} \leqslant \alpha \leqslant \frac{3}{4}$. The sum in (28.29) is thus

$$\ll h\lambda^{28} \max_{|\gamma| < T} (xT^{-12/5})^{\beta - 1}, \tag{28.31}$$

the maximum being over zeros ρ with $|\gamma| < T$. With

$$T = x^{5/12 - \delta}, \tag{28.32}$$

where $\delta > 0$, the expression in (28.31) is

$$\ll h\lambda^{28} \max_{|\gamma| < T} x^{\delta(\beta - 1)} \ll h\lambda^{28} \exp\{\delta\lambda(A\lambda^{-B})\}, \tag{28.33}$$

which is $o(1)$ as x tends to infinity; here we have used (28.22). The error term in (28.27) is also less than h when

$$h \gg x^{7/12 + \delta} \lambda^2. \tag{28.34}$$

If (28.34) holds with a sufficiently large constant then

$$\psi(x+h) - \psi(x) > \{\tfrac{1}{2} - o(1)\}h. \tag{28.35}$$

Finally we note that the prime powers up to $x+h$ contribute

$$\ll x^{\frac{1}{2}}\lambda^2 \tag{28.36}$$

to the sum $\psi(x+h)-\psi(x)$, and so for sufficiently large x

$$\sum_{x<p\leqslant x+h} \log p \geqslant \tfrac{1}{4}h, \tag{28.37}$$

and we have proved (28.20) when we choose

$$\delta = \tfrac{1}{2}(c-\tfrac{7}{12}). \tag{28.38}$$

NOTATION

$\sum_p, \prod_p$	indicate a sum or a product over primes only (Chapter 1)
(m, n)	highest common factor of the integers m and n (Chapter 1)
$m \equiv n \pmod q$	$m-n$ is a multiple of q (Chapter 1)
$e(\alpha) = \exp 2\pi i\alpha$, $e_q(\alpha) = \exp(2\pi i\alpha/q)$	(1.4)
$\sum_{a \bmod q}$	sum over a set of representatives of residue classes mod q (Chapter 1)
$\sum^*_{a \bmod q}$	sum over a set of representatives of reduced residue classes mod q (Chapter 1)
$\varphi(m)$	Euler's function (1.10)
$c_q(m)$	Ramanujan's sum (1.11)
$s = \sigma+it$	complex variable (1.16)
$\chi(m)$	a Dirichlet's character (Chapter 1)
$d(m)$	the number of divisors of m (1.23)
$\mu(m)$	Möbius's function (1.25)
$\Lambda(m)$	$\log p$ if m is a prime power p^a, otherwise 0 (1.31)
$\psi(x)$	sum function of $\Lambda(m)$ (2.1)
$f(x) \ll g(x)$	$\|f(x)\| = O(g(x))$ (2.3)
$f(\chi)$	the conductor of a character (Chapter 3)
$\tau(\chi)$	Gauss's sum (3.7)
$\chi_0(m)$	a trivial character (Chapter 3)
$[\alpha]$	largest integer not exceeding α (4.3)
$\\|\alpha\\|$	distance of α from the nearest integer (4.4)
$H(\alpha)$	the saw-tooth Fourier series (4.5)
$L(s, \chi)$	Dirichlet's L-function (5.8)
$\zeta(s)$	Riemann's zeta function (5.9)
$S(\alpha)$	an exponential sum (Chapter 6)
$\pi(x)$	number of primes up to x (Chapter 6)
a/q	a rational number in its lowest terms (Chapter 6)
$\Gamma(s)$	Euler's gamma function (11.1), (11.12)
$\xi(s)$, $\xi(s, \chi)$	functions occurring in function equations for $\zeta(s)$ and $L(s, \chi)$ (12.2), (12.3)
$\rho = \beta+i\gamma$	a zero of $\xi(s)$ or of $\xi(s, \chi)$ (Chapter 12)
$\psi(x, \chi)$	sum function of $\Lambda(m)\chi(m)$ (17.1)
$\psi(x; q, a)$	sum function of $\Lambda(m)$ in the arithmetic progression $a \pmod q$ (17.18)
S_χ	a character sum corresponding to the exponential sum $S(\alpha)$ (18.4)
$\sum^*_{\chi \bmod q}$	a sum over proper characters mod q (Chapter 18)
$u = \lambda+i\tau$	auxiliary complex variable (Chapter 20)
$G(u)$	all the 'junk' in the functional equation (20.2)
$M(s, \chi)$	partial sum for the inverse of $L(s, \chi)$ (23.2)
$N(\chi)$	number of zeros of $L(s, \chi)$ in a rectangle (Chapter 23)

BIBLIOGRAPHY

The epigraphs are from
I. *Winnie the Pooh*
II. *The House at Pooh Corner*

BOMBIERI, E. (1965). On the large sieve. *Mathematika* **12,** 201–25.

—— (1972). A note on the large sieve. To appear.

—— and DAVENPORT, H. (1968). On the large sieve method. *Abhandlungen aus Zahlentheorie und Analysis zur Erinnerung an Edmund Landau.* Berlin.

—— —— (1969). Some inequalities involving trigonometric polynomials. *Annali Scu. norm. sup., Pisa* **23,** part 2, 223–41.

CHANDRASEKHARAN, K. and NARASIMHAN, R. (1963). The approximate functional equation for a class of zeta-functions. *Math. Annaln* **152,** 30–64.

CRAMÉR, H. (1936). On the order of magnitude of the difference between consecutive prime numbers. *Acta arith.* **2,** 23–46.

DAVENPORT, H. (1967). *Multiplicative number theory.* Markham, Chicago.

ESTERMANN, T. (1948). On Dirichlet's L-functions. *J. Lond. math. Soc.* **23,** 275–9.

FOGELS, E. (1969). Approximate functional equation for Hecke's L-functions of quadratic field. *Acta arith.* **16,** 161–78.

FRANEL, J. (1924). Les suites de Farey et les problèmes des nombres premiers. *Nachr. Ges. Wiss. Göttingen,* 198–201.

GALLAGHER, P. X. (1967). The large sieve. *Mathematika* **14,** 14–20.

—— (1968). Bombieri's mean value theorem. Ibid. **15,** 1–6.

HALÁSZ, G. and TURÁN, P. (1969). On the distribution of the roots of Riemann Zeta and allied functions I. *J. Number Theory* **1,** 121–37.

HALBERSTAM, H. and ROTH, K. F. (1966). *Sequences,* vol. 1. Oxford.

HARDY, G. H. and WRIGHT, E. M. (1960). *An introduction to the theory of numbers.* 4th edn. Oxford.

HOHEISEL, G. (1930). Primzahlprobleme in der Analysis. *Sber. berl. math. Ges.* 580–8.

INGHAM, A. E. (1937). On the difference between consecutive primes. *Q. Jl Math.* **8,** 255–66.

—— (1940). On the estimation of $N(\sigma, T)$. Ibid. **11,** 291–2.

JEFFREYS, H. and JEFFREYS, B. (1962). *Methods of mathematical physics.* Cambridge.

LANDAU, E. (1927). *Vorlesungen über Zahlentheorie.* Leipzig.

LINNIK, YU. V. (1945). On the possibility of a unique method in certain problems of additive and multiplicative number theory. *Doklady Akad. Nauk SSSR, ser. mat.* **49,** 3–7.

—— (1964). All large numbers are sums of a prime and two squares (a problem of Hardy and Littlewood) II. *Am. math. Soc. Transl.* (2) **37,** 197–240.

—— and RÉNYI, A. (1947). On some hypotheses in the theory of Dirichlet characters. *Izv. Akad. Nauk SSSR, ser. mat.* **11,** 539–46.

MONTGOMERY, H. L. (1968). A note on the large sieve. *J. Lond. math. Soc.* **43,** 93–8.

—— (1969*a*). Mean and large values of Dirichlet polynomials. *Invent. math.* **8,** 334–45.

—— (1969*b*). Zeros of L-functions. Ibid. **8,** 346–54.

—— (1971). *Lectures on multiplicative number theory.* Springer.

PÓLYA, G. (1918). Über die Verteilung der quadratischen Reste und Nichtreste. *Nachr. Ges. Wiss. Göttingen,* 21–9.

PRACHAR, K. (1957). *Primzahlverteilung.* Springer.

ROTH, K. F. (1965). On the large sieves of Linnik and Rényi. *Mathematika* **12,** 1–9.

TITCHMARSH, E. C. (1951). *The theory of the Riemann zeta-function.* Oxford.

VINOGRADOV, A. I. (1965). On the density hypothesis for Dirichlet L-functions. *Izv. Akad. Nauk SSSR, ser. mat.* **29,** 903–34.

VINOGRADOV, I. M. (1954). *The method of trigonometric sums in the theory of numbers* (transl. A. Davenport and K. F. Roth). Interscience, New York.

—— (1955). *An introduction to the theory of numbers* (translation). Pergamon, Oxford.

INDEX